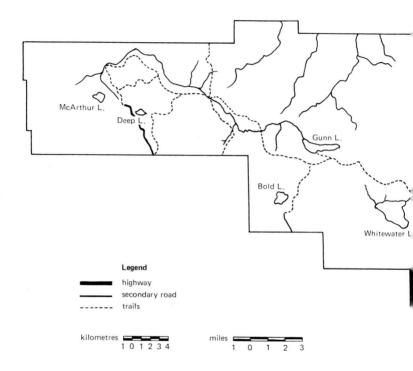

McArthur L.

Deep L.

Gunn L.

Bold L.

Whitewater L

Legend

▬▬▬ highway

───── secondary road

------- trails

kilometres ▭▬▭▬▭
1 0 1 2 3 4

miles ▭▬▭▬▭
1 0 1 2 3

Riding Mountain National Park, Manitoba

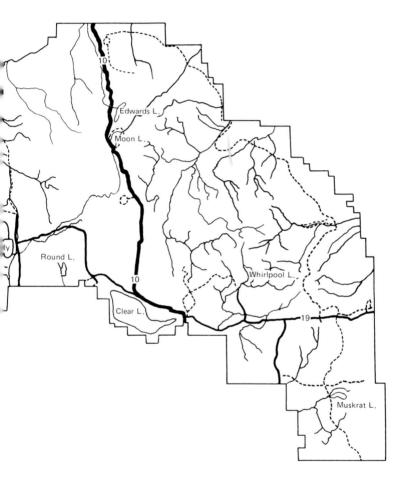

Plants of Riding Mountain National Park, Manitoba

William J. Cody, Curator
Vascular Plant Herbarium
Biosystematics Research Centre

Research Branch
Agriculture Canada

Publication 1818/E
1988

Published in cooperation with the Canadian Parks Service, Environment Canada.

©Minister of Supply and Services Canada 1988
Available in Canada through
Authorized Bookstore Agents
and other bookstores
or by mail from
Canadian Government Publishing Centre
Supply and Services Canada
Ottawa, Canada K1A 0S9

Catalogue No. A53-1818/1988E
ISBN 0-660-12879-9

Price subject to change without notice.

Canadian Cataloguing in Publication Data

Cody, William J., 1922–

   Plants of Riding Mountain National Park,
Manitoba

(Publication ; 1818E)

Issued also in French under title: Flore du Parc
national du mont Riding, Manitoba.
Includes index.
Bibliography: p.

1. Botany--Manitoba--Riding Mountain National
Park. I. Canada. Agriculture Canada.
II. Canadian Parks Service. III. Title.
IV. Series: Publication (Canada. Agriculture
Canada). English ; 1818E.

QK203.M3.C6 1988   581.97127'2   C88-099202-6

Cover illustration: *Rudbeckia laciniata* L.; tall coneflower
(photograph by author).

Staff Editor
Frances Smith

Contents

Introduction

Riding Mountain, like Turtle Mountain, Duck Mountain, and Porcupine Mountain, forms a part of the Manitoba Escarpment. It stands out on the otherwise relatively flat countryside of southern Manitoba. The sharpest feature is the steep escarpment on the eastern flank which rises to a height of about 400 m. To the west of the escarpment is a rolling plateau.

Cretaceous shales are exposed and deeply incised by streams on the escarpment. On the plateau, the surface deposits are mainly glacial tills, but some lacustrine materials can be found around the lakes. Where drainage is poor, shallow peat deposits occur.

Most of Riding Mountain National Park lies in the Mixedwood Section of the Boreal Forest Region (Rowe 1959). This is characterized by *Populus tremuloides* (aspen poplar), *P. balsamifera* (balsam poplar), *Betula papyrifera* (white birch, paper birch), *Picea glauca* (white spruce), and *Abies balsamea* (balsam fir) on well-drained sites; *Pinus banksiana* (jack pine) on drier sites; and *Picea mariana* (black spruce) and *Larix laricina* (tamarack) in low, poorly drained situations, although *Pinus banksiana* and *Picea mariana* do grow together on a few well to moderately drained slopes. Also present in the park are the broad-leaved trees *Ulmus americana* (American or white elm), *Fraxinus pennsylvanica* (green ash), *Acer negundo* (Manitoba maple), and *Quercus macrocarpa* (bur oak). These broad-leaved trees are found largely on or below the escarpment, where they are associated with such rare plants of eastern affinity as *Celastrus scandens* (bittersweet), *Parthenocissus inserta* (Virginia creeper), and *Amphicarpa bracteata* (hog-peanut). They form an intrusion of the Aspen–oak section of the Boreal forest region (Rowe 1959).

Although most of Riding Mountain is treed, areas of Rough Fescue Grassland (dominated by *Festuca hallii*) and Mixed Grassland occur, particularly in the western region. Notable examples are in the area of the bison enclosure and Birdtail Valley. Diagram 1 depicts the relationship of tree species to edaphic and physiographic factors in the Park, and Diagram 2 shows a generalized pattern of the distribution of major plant communities.

Within the boundaries of Riding Mountain National Park, a total of 88 families, which include 300 genera, 669 species, and two hybrids, are known to occur. This publication is intended to provide a workable key to the vascular plants found within Riding Mountain National Park. Although the descriptions are brief, they are, it is hoped, complete enough to separate the various species from one another. The book is based on studies carried out by the author during the summers of 1979 and 1983 and on the study of extensive collections made during the first year, in particular. The herbaria of the Canada Department of Agriculture (DAO) and the National Museum of Natural Sciences (CAN), both in Ottawa, were surveyed, as were those of the University of Manitoba (MAN), Winnipeg, Canadian Wildlife Service and Canadian Forestry Service (CAFB), Edmonton, and Riding Mountain National Park. Some material preserved in the herbarium of the Manitoba Museum of Man and Nature, Winnipeg, was also examined. Earlier botanical works such as Scoggan (1957), Lowe (1943), and Boivin (1967–1981) were also checked.

The illustrations are reproduced from *Vascular Plants of Continental Northwest Territories, Canada,* by A.E. Porsild and W.J. Cody (1980), with permission from the National Museum of Natural Sciences, National Museums of Canada. Diagrams 1 and 2 are reproduced from *Notes on the Vegetation in Riding Mountain National Park, Manitoba* by R.H. Bailey (1968), with permission from Environment Canada.

The curators of the herbaria mentioned above are gratefully acknowledged for making specimens available for study. W.A. Wojtas provided technical assistance in the field and in the herbarium, and B.S. Brooks provided help in the preparation of the text. The manuscript was reviewed by I.J. Bassett, V.L. Harms, J.D. Johnson, E. Small, and G. Trottier, whose comments were much appreciated.

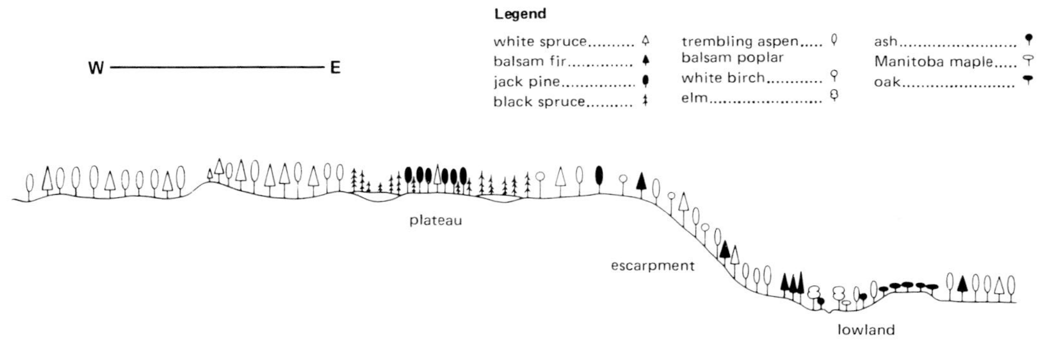

Parent material	glacial till	coarse outwash	glacial till	peat	glacial till	peat	glacial till	recent alluvium	gravel beach	glacial till
Soil	gray wooded	dark gray wooded	gray wooded	organic	gray wooded	organic	gray wooded	regosol	black	gray wooded
Soil texture	loam to clay loam	coarse sand to sandy loam	loam to clay loam	fibrous	very fine sandy loam to clay loam	fibrous	very fine sandy loam to clay	silty clay loam	loamy sand	fine sandy loam to clay loam
Soil moisture	fresh to very moist	dry	fresh to very moist	wet	dry to fresh	wet	fresh to very moist	very moist	dry	fresh to very moist

Diagram 1. Relationship of tree species to edaphic and physiographic factors in Riding Mountain National Park.

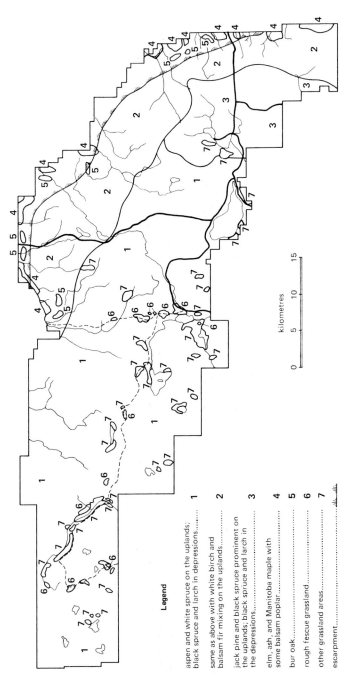

Legend

aspen and white spruce on the uplands;
black spruce and larch in depressions...... 1

same as above with white birch and
balsam fir mixing on the uplands...... 2

jack pine and black spruce prominent on
the uplands; black spruce and larch in
the depressions...... 3

elm, ash, and Manitoba maple with
some balsam poplar...... 4

bur oak...... 5

rough fescue grassland...... 6

other grassland areas...... 7

escarpment...... ⊥⊥⊥

kilometres

0   5   10   15

Diagram 2.   A generalized pattern of the distribution of major plant communities (after Bailey 1968)
in Riding Mountain National Park.

4

Key to the families

1a. Plants without flowers or seeds; reproducing by spores
.... (2)
1b. Plants producing seeds .... (7)

2a. Leaves slender, often scale-like, simple, sessile, mostly small .... (3)
2b. Leaves broad, usually more than 2 cm long, often quite large, variously incised or dissected .... (5)

3a. Stems not conspicuously jointed; leaves mostly imbricated; spore cases in cone-like spikes (strobiles)
.... (4)
3b. Stems conspicuously jointed, mostly hollow; leaves scale-like, in sheath-like whorls at the nodes; spore cases on the scales of terminal cone-like spikes (strobiles) .... 3. EQUISETACEAE p. 20

4a. Leaves without a ligule; strobiles terete; homosporous
.... 1. LYCOPODIACEAE p. 19
4b. Leaves ligulate; strobiles 4-sided; sporangia of 2 kinds, microsporangia containing many minute microspores (male) and macrosporangia containing fewer and larger macrospores (female) ....
.... 2. SELAGINELLACEAE p. 19

5a. Spore cases relatively large, borne in a terminal (grape-like) cluster, the sterile blade appearing lateral on a common stalk with it ....
.... 4. OPHIOGLOSSACEAE p. 24
5b. Spore cases minute, borne in clusters (sori) on the back or near the margins of green blades or on separate modified fronds .... (6)

6a. Fronds robust, 3-forking below the blade-bearing portion, scattered from deeply buried thick rhizomes; sori marginal .... 5. PTERIDACEAE p. 26
6b. Fronds usually delicate, pinnate or ternate, densely tufted or scattered along a thin rhizome ....
.... 6. ASPIDIACEAE p. 26

7a. Trees or shrubs with needle-like or scale-like leaves; leaves not falling in autumn (except *Larix*); seeds produced directly on the scales of cones (the cones sometimes berry-like) ...... 7. PINACEAE p. 30

7b. Woody or nonwoody plants with variously shaped usually deciduous leaves; seeds produced in various types of fruit; fruit usually not in cone-like structures .... (8)

8a. Leaves usually parallel-veined; parts of flowers usually in threes or sixes, never in fives; herbs ... (9)

8b. Leaves usually net-veined; parts of flowers usually in fours or fives; herbs, shrubs, or trees .... (23)

9a. Plants not differentiated into stem and leaf, small, thallus-like, ellipsoid, oblong or globose; free-floating, immersed, or sometimes stranded aquatics .... 18. LEMNACEAE p. 100

9b. Plants with stems and leaves, not thallus-like (10)

10a. Perianth lacking or inconspicuous, often consisting of bristles or scales, not petal-like (see also Scheuchzeriaceae and Juncaceae) .... (11)

10b. Perianth of 2 distinct whorls, with the inner often petal-like and conspicuous .... (17)

11a. Flowers enclosed or subtended by scales (glumes); plants grass-like, with jointed stems, sheathing leaves and 1-seeded fruit .... (12)

11b. Flowers not enclosed in scales (though sometimes in involucrate heads); plants not as above .... (13)

12a. Stems usually hollow, terete, or flattened; leaves 2-ranked; leaf sheaths usually split (open); anthers attached at the middle ... 15. GRAMINEAE p. 42

12b. Stems solid, usually more or less 3-sided; leaves usually 3-ranked; leaf-sheaths not split (closed); anthers attached at the base .... 16. CYPERACEAE p. 74

13a. Aquatic plants; stems and leaves flaccid, either immersed or floating (see also *Sparganium*) .... (14)

13b. Terrestrial plants or bases in water; stems rigid enough to support shoots above water level .... (15)

14a. Leaves alternate or subopposite ....
.... 10. POTAMOGETONACEAE p. 34
14b. Leaves opposite or whorled ....
.... 11. NAJADACEAE p. 38

15a. Flowers in globose heads; perianth of flat scales ....
.... 9. SPARGANIACEAE p. 34
15b. Flowers mostly in spikes .... (16)

16a. Flowers unisexual, with the pistillate in a thick spike
and the staminate above ... 8. TYPHACEAE p. 32
16b. Flowers perfect, in a thick dense fleshy spike, with the
spike subtended by a large bract (spathe) ....
.... 17. ARACEAE p. 100

17a. Plants aquatic, immersed or nearly so; leaves whorled
.... 14. HYDROCHARITACEAE p. 42
17b. Plants terrestrial or semi-aquatic; leaves not whorled
.... (18)

18a. Perianth relatively inconspicuous, green or brownish;
plants rush-like .... (19)
18b. Perianth conspicuous, at least the inner whorl
brightly colored .... (20)

19a. Perianth dry, often scarious; flowers commonly in
panicles or heads; carpels 3, united, forming a small
capsule .... 19. JUNCACEAE p. 102
19b. Perianth herbaceous; flowers in racemes or spikes;
carpels 3 or 6, almost distinct, separating as follicles
when ripe ..... 12. SCHEUCHZERIACEAE p. 38

20a. Carpels numerous, distinct, in a ring or a cluster,
becoming achenes; plants of marshes and bogs .....
.... 13. ALISMATACEAE p. 40
20b. Carpels 3; ovaries united; fruit a capsule or berry
.... (21)

21a. Ovary superior (rarely partly inferior) ....
.... 20. LILIACEAE p. 106
21b. Ovary inferior .... (22)

22a. Flowers regular; stamens 3 ....
.... 21. IRIDACEAE p. 110
22b. Flowers very irregular; stamens 1 or 2 ....
.... 22. ORCHIDACEAE p. 110

23a. Corolla none; calyx present or absent .... (24)
23b. Corolla and calyx both present .... (52)

24a. Flowers unisexual, either staminate or pistillate
.... (25)
24b. Flowers with both stamens and pistil .... (41)

25a. Trees or shrubs .... (26)
25b. Herbs .... (32)

26a. Leaves pinnately compound .... (27)
26b. Leaves simple .... (28)

27a. Leaflets mostly 3–5, entire, or coarsely few-toothed, or
lobed; fruit 2 separable inequilateral samaras .....
.... 53. ACERACEAE p. 189
27b. Leaflets 5–7, more regularly and finely toothed, not
lobed; fruit a single equilateral samara ....
.... 70. OLEACEAE p. 209

28a. Leaves linear, evergreen, soon reflexed, 2.5–7.0 mm
long .... 50. EMPETRACEAE p. 188
28b. Leaves dilated, longer, spreading or ascending
.... (29)

29a. Leaves opposite, scurfy with rusty scales beneath or
silvery on both sides ....
.... 60. ELAEAGNACEAE p. 194
29b. Leaves alternate, scurfy scales lacking .... (30)

30a. Fruit a many-seeded capsule; seeds furnished with
long silky down .... 23. SALICACEAE p. 116
30b. Fruit a nut or nutlet   (31)

31a. Fruit an acorn; leaves deeply lobed ....
.... 25. FAGACEAE p. 128
31b. Fruit a wingless nut enclosed in a leafy involucre or
winged nutlets in a bracted catkin ....
.... 24. BETULACEAE p. 126

32a. Immersed aquatics, rooting in the mud; upper leaves
sometimes floating on the surface .... (33)
32b. Terrestrial .... (34)

33a. Leaves whorled, finely dissected into capillary to
linear serrate divisions ....
.... 35. CERATOPHYLLACEAE p. 144

33b. Leaves opposite, linear to obovate, entire ....
.... 49. CALLITRICHACEAE p. 188

34a. Leaves 2–3 – ternately compound ....
.... 37. RANUNCULACEAE p. 146
34b. Leaves simple .... (35)

35a. Nodes of stem and panicled racemes covered by
tubular sheaths (ocreae) ....
.... 30. POLYGONACEAE p. 130
35b. Nodes without tubular sheaths .... (36)

36a. Leaves cordate, mostly 3–5–(7)-lobed or those of the
branches sometimes uncleft; plant twining, harshly
scabrous .... 27. CANNABACEAE p. 128
36b. Leaves not obviously lobed; plants not twining (37)

37a. Plants with milky juice ....
.... 48. EUPHORBIACEAE p. 188
37b. Plants with clear juice .... (38)

38a. Leaves opposite, petioled; stinging bristles present
.... 28. URTICACEAE p. 130
38b. Leaves alternate; stinging bristles absent ..... (39)

39a. Spikes or heads of flowers close or continuous ....
.... 31. CHENOPODIACEAE p. 135
39b. Spikes or heads of flowers interrupted .... (40)

40a. Flowers bracted at the base; bracts and sepals scarious
.... 32. AMARANTHACEAE p. 138
40b. Flowers bractless; sepals herbaceous or fleshy ....
.... 31. CHENOPODIACEAE p. 135

41a. Trees or shrubs .... (42)
41b. Herbs .... (45)

42a. Leaves silvery-scurfy on both sides ....
.... 60. ELAEAGNACEAE p. 194
42b. Leaves not scurfy .... (43)

43a. Leaves opposite, palmately veined ....
.... 53. ACERACEAE p. 189
43b. Leaves alternate, pinnately veined .... (44)

55a. Plants climbing by tendrils; leaves digitate ....
.... 56. VITACEAE p. 190
55b. Plants climbing by twining; leaves simple ....
.... 52. CELASTRACEAE p. 189

56a. Leaves compound, alternate .... (57)
56b. Leaves simple .... (59)

57a. Petals 5, irregular; stamens 10, united at the very
base; fruit a legume .. 43. LEGUMINOSAE p. 176
57b. Petals 5, all the same; stamens distinct to the base;
fruit not a legume .... (58)

58a. Stamens numerous; flowers perfect ....
.... 42. ROSACEAE p. 166
58b. Stamens 5; flowers dioecious or polygamous ....
.... 51. ANACARDIACEAE p. 189

59a. Fruit 2 separable 1-seeded samaras, or "keys"; leaves
opposite, palmately veined ....
.... 53. ACERACEAE p. 189
59b. Fruit not a samara; leaves opposite or alternate,
pinnately veined .... (60)

60a. Anthers opening by apical pores ....
.... 68. ERICACEAE p. 204
60b. Anthers not opening by apical pores .... (61)

61a. Leaves opposite or clustered toward the ends of the
twigs .... 66. CORNACEAE p. 202
61b. Leaves alternate .... (62)

62a. Stamens 5; fruit a many-seeded berry ....
.... 41. SAXIFRAGACEAE p. 162
62b. Stamens numerous ...... 42. ROSACEAE p. 166

63a. Plants aquatic; leaves submersed or floating on the
surface .... (64)
63b. Plants terrestrial, of dry to swampy or muddy habitats
.... (66)

64a. Leaves mostly submersed, coarsely to finely divided,
covering the stem .... (65)
64b. Leaves mostly floating, suborbicular to reniform,
entire except for the basal sinus ....
.... 36. NYMPHAEACEAE p. 146

11

65a. Flowers unisexual, sessile in the axils of entire or pinnate bracts near the summit of the stem; stamens 8 .... 62. HALORAGACEAE p. 196
65b. Flowers perfect, pedicelled; stamens usually more numerous .... 37. RANUNCULACEAE p. 146

66a. Anthers opening by apical pores .... 67. PYROLACEAE p. 202
66b. Anthers not opening by apical pores .... (67)

67a. Ovary inferior or appearing so, adnate to or enclosed by the calyx tube .... (68)
67b. Ovary superior, not adnate to or enclosed by the calyx tube .... (72)

68a. Leaves simple .... (69)
68b. Leaves compound .... (70)

69a. Flowers in a dense head-like cyme subtended by 4 broad white or purple-tipped petaloid bracts; fruit a red drupe with a 1- or 2-seeded stone .... 66. CORNACEAE p. 202
69b. Flowers neither cymose nor subtended by petaloid bracts; fruit a capsule .. 61. ONAGRACEAE p. 194

70a. Inflorescence a spike-like raceme of yellow flowers; throat of calyx with hooked bristles .... 42. ROSACEAE p. 166
70b. Inflorescence a simple or compound umbel ..... (71)

71a. Fruit consisting of 2 seed-like, dry, 1-seeded carpels cohering by their inner face and separating later; styles 2 .... 65. UMBELLIFERAE p. 198
71b. Fruit a 5-seeded blackish drupe; styles usually more than 2 .... 64. ARALIACEAE p. 198

72a. Plants insectivorous, occurring in boggy habitats; leaves all basal, covered with gland-tipped hairs ... . 40. DROSERACEAE p. 162
72b. Plants not insectivorous, mostly of drier habitats .... (73)

73a. Flowers irregular .... (74)
73b. Flowers regular or nearly so .... (76)

74a. Leaves compound, much dissected, glaucous; flowers golden yellow .... 38. FUMARIACEAE p. 154
74b. Leaves entire or deeply divided to the base into narrow segments .... (75)

75a. Tall plants with watery juice; petals 2, 2-lobed; sepals 4, one spurred at the base ....
.................54. BALSAMINACEAE p. 190
75b. Plants short, not juicy; petals 5, the lower one spurred at the base; sepals 5, not spurred ....
.... 59. VIOLACEAE p. 192

76a. Stamens borne on the receptacle .... (77)
76b. Stamens adnate to base of perianth, or inserted on a disk or thickened zone beneath the ovary ...... (80)

77a. Sepals or calyx lobes 2 ....
.... 33. PORTULACACEAE p. 139
77b. Sepals or calyx lobes 4–7 .... (78)

78a. Stamens 6 (4 long and 2 short); sepals and petals 4 .... 39. CRUCIFERAE p. 154
78b. Stamens 4 to many; sepals and petals mostly 5 (79)

79a. Pistil solitary; fruit a capsule; stamens 4–10; leaves simple, mostly entire and opposite ....
.... 34. CARYOPHYLLACEAE p. 140
79b. Pistils few to many, distinct; fruit either capsules, achenes, or berries; stamens often more than 10; leaves simple or compound, entire or toothed, or lobed, mostly alternate or basal ....
.... 37. RANUNCULACEAE p. 146

80a. Leaves compound ....(81)
80b. Leaves simple, entire to deeply cleft .... (82)

81a. Leaflets 3, obcordate .. 45. OXALIDACEAE p. 186
81b. Leaflets 3 or more, but not obcordate ....
.....................42. ROSACEAE p. 166

82a. Stamens numerous, united into a central column around the pistil; leaves alternate, shallowly lobed to deeply cleft .... 57. MALVACEAE p. 190
82b. Stamens 4 to many, distinct or united only at the base .... (83)

83a. Leaves mostly opposite .... (84)
83b. Leaves mostly alternate or basal .... (85)

13

84a. Leaves entire, sessile or nearly so, dotted; stamens 9 to many, often in 3–5 clusters ....
.... 58. HYPERICACEAE p. 191
84b. Leaves toothed to deeply cleft; stamens 5–10, not in clusters .... 46. GERANIACEAE p. 186

85a. Stamens 5 or 10 .... (86)
85b. Stamens numerous ...... 42. ROSACEAE p. 166

86a. Stamens 5, their filaments united at the base; leaves linear to linear-lanceolate, entire ....
.... 44. LINACEAE p. 186
86b. Stamens 10, distinct (if 5, the leaves broader and chiefly basal) ..... 41. SAXIFRAGACEAE p. 162

87a. Stamens more numerous than the lobes of the corolla .... (88)
87b. Stamens not more numerous than the lobes of the corolla .... (92)

88a. Leaves simple, entire to deeply lobed .... (89)
88b. Leaves compound (at least the lower ones) ..... (91)

89a. Flowers very irregular; petals 3, connected with the stamen tube; leaves entire or finely toothed ....
.... 47. POLYGALACEAE p. 186
89b. Flowers regular or nearly so .... (90)

90a. Stamens numerous, united into a central column around the pistil .... 57. MALVACEAE p. 190
90b. Stamens 8 or 10, distinct; leaves entire or finely toothed .... 68. ERICACEAE p. 204

91a. Sepals 2; petals in 2 pairs; stamens in 2 sets of 3 each; filaments often united; leaves exstipulate; leaflets finely dissected ..... 38. FUMARIACEAE p. 154
91b. Sepals united; the calyx tube 4- or 5-toothed; petals usually 5; corolla more or less distinctly papilionaceous; stamens 10, 9 or all of them united into a tube; leaves stipulate; leaflets entire or toothed .... 43. LEGUMINOSAE p. 176

92a. Shrubs, climbers, or trailing plants; stamens alternate with the corolla lobes or fewer in number ...... (93)
92b. Herbs .... (96)

93a. Plants climbing or evergreen and trailing ..... (94)
93b. Shrubs .... (95)

94a.   Plants climbing; leaves alternate, broad, and green or, in *Cuscuta*, reduced to a few minute scales .... 74. CONVOLVULACEAE   p. 212

94b.   Plants trailing; leaves opposite .... 84. CAPRIFOLIACEAE   p. 226

95a.   Ovary superior, free from the calyx tube; anthers upright, opening by terminal pores .... 68. ERICACEAE   p. 204

95b.   Ovary inferior, adnate to the calyx tube; anthers not opening by terminal pores .... 84. CAPRIFOLIACEAE   p. 226

96a.   Stamens of the same number as corolla lobes and opposite them .... 69. PRIMULACEAE   p. 208

96b.   Stamens alterate with the corolla lobes or fewer (96)

97a.   Ovary inferior, adherent to the calyx tube ..... (98)

97b.   Ovary superior, free from the calyx tube ..... (102)

98a.   Flowers crowded in a dense head on a common receptacle, surrounded by an involucre; fruit a dry seed-like achene ...... 88. COMPOSITAE   p. 231

98b.   Flowers not in dense heads .... (99)

99a.   Leaves alternate; stamens 5; anthers free or united into a tube .... (100)

99b.   Leaves opposite or whorled; anthers free ..... (101)

100a.   Corolla regular; anthers separate .... 86. CAMPANULACEAE   p. 230

100b.   Corolla irregular; anthers united .... 87. LOBELIACEAE   p. 230

101a.   Stamens 3, always fewer than the corolla lobes; calyx lobes becoming pappus-like; leaves opposite .... 85. VALERIANACEAE   p. 230

101b.   Stamens 4 or 5; calyx lobes (when present) not pappus-like; leaves opposite or whorled .... 83. RUBIACEAE   p. 224

102a.   Corolla irregular .... (103)

102b.   Corolla regular or nearly so .... (105)

103a. Stem square; leaves opposite; ovary 4-lobed or 4-parted, with each lobe forming a seed-like nutlet or achene at the base of the style; stamens a single pair or 2 pairs of unequal length ... 78. LABIATAE p. 215

103b. Stem round; leaves alternate, opposite, or whorled; ovary unlobed, forming a usually several to many-seeded capsule tipped by the style ..... (104)

104a. Plants insectivorous, aquatic (with bladder traps borne on branches or on finely dissected submersed leaves) or terrestrial (with rosettes of entire broad leaves); stamens 2; capsule 1-locular .... 81. LENTIBULARIACEAE p. 223

104b. Plants not insectivorous, occurring in wet or dry habitats; stamens 2 or 2 pairs of unequal length; capsule 2-locular .... 80. SCROPHULARIACEAE p. 220

105a. Stamens fewer than the corolla lobes .... (106)
105b. Stamens as many as the corolla lobes .... (107)

106a. Leaves in a basal rosette; fruit a 2-locular capsule, with the top falling off like a lid .... 82. PLANTAGINACEAE p. 224

106b. Leaves opposite on the square stem; fruit consisting of 4 seed-like nutlets or achenes at the base of the style .... 78. LABIATAE p. 215

107a. Stem square .... 78. LABIATAE p. 215
107b. Stem roundish .... (108)

108a. Ovary deeply 4-lobed, forming 4 seed-like nutlets around the base of the style; leaves alternate, with the surface often very rough to the touch .... 77. BORAGINACEAE p. 214

108b. Ovary unlobed; leaves smoother .... (109)

109a. Plants with an acrid milky juice; fruit a many-seeded follicle; seeds with a coma .... (110)

109b. Plants without milky juice; fruit a capsule or berry .... (111)

110a. Flowers in umbels; follicle ovoid or thick-lanceolate .... 73. ASCLEPIADACEAE p. 211

110b. Flowers in clusters at the end of branches and in axils of leaves; follicle long and narrow .... 72. APOCYNACEAE p. 211

111a.   Fruit a berry; anthers separate or forming a tube around the style ...... 79. SOLANACEAE p. 220
111b   Fruit a capsule; anthers distinct .... (112)

112a.   Ovary and capsule 1-locular .... (113)
112b.   Ovary and capsule 2- or 3-locular; leaves simple ....
.... (114)

113a.   Leaves simple, entire, opposite, or whorled (or 3-foliate and alternate in *Menyanthes*); style not divided ....
.... 71. GENTIANACEAE p. 210
113b.   Leaves alternate, pinnatifid, but not 3-foliate; style 2-cleft .... 76. HYDROPHYLLACEAE p. 214

114a.   Leaves alternate, petioled, at least the upper ones narrowly to broadly oval, cordate or subtruncate to tapering at the base; flowers large, funnelform, solitary in the axils ....
.... 74. CONVOLVULACEAE p. 212
114b.   Leaves alternate or opposite, sessile, linear to oblong-lanceolate, rounded to tapering at the base; flowers smaller, cymose or clustered ....
.... 75. POLEMONIACEAE p. 212

The flora

1. LYCOPODIACEAE   club-moss family

Lycopodium   **club-moss**

1a.   Sporangia in the axils of leaf-like sporophylls; few-forked leafy stems ascending and sprawling, rooting toward the base from among the brown marcescent leaves; leaves oblanceolate, spreading or reflexed, acuminate, erose-serrulate near the apex. *Lycopodium lucidulum* Michx.; shining club-moss. Rare on moss-covered shale under birch near the East Gate.

1b.   Sporangia in the axils of modified terminal leafy-bracted strobiles .... (2)

2a.   Strobili sessile .... (3)

2b.   Strobili peduncled .... (4)

3a.   Aerial stems arising from elongate prostrate stems; leaves more or less stiff and hard, linear-subulate to linear-oblanceolate, tipped by a sharp spinule. *Lycopodium annotinum* L.; bristly club-moss; Fig. 1. Rare in rich moist woodland.

3b.   Aerial stems erect and tree-like, arising singly from the buried rhizome-like horizontal stems; leaves divergent, strongly decurrent, with the free part linear-attenuate. *Lycopodium dendroideum* Michx. (*L. obscurum* pro parte); round-branched ground-pine. Under dense shrubs on the steep east slope; localized.

4a.   Leaves with long hair-like tips, linear-subulate; horizontal stems elongated; leaves uniform, but the lower turned upward, rooting at intervals; erect branches at first simple, becoming dichotomous; fertile branch with a leafy-bracted peduncle. *Lycopodium clavatum* L. var. *monostachyon* Hook. & Grev.; common club-moss; Fig. 2. Rare in jack pine woodland.

4b.   Leaves scale-like; stems horizontal, mostly below the surface of the ground; upright stems with crowded or somewhat forking branchlets; branchlets flattened, often strongly constricted between yearly growths; strobili mostly 1 or 2 on remotely bracted peduncles. *Lycopodium complanatum* L.; flatbranch club-moss; Fig. 3. Rare in jack pine woodland.

2. SELAGINELLACEAE   spike-moss family

Selaginella   **spike-moss**

1a.   Delicate branching plants forming small mats; leaves uniform 2–4 mm long, spreading-ascending, acute, ciliate; fertile branches upright, with the lower leaves similar to those of the stem but becoming larger upwards to form the sporophylls of a subcylindric spike. *Selaginella selaginoides* (L.) Link; Fig. 4. Under black spruce in moss at border of marl bog; rare.

1b.   Stiff evergreen plants forming compact carpets; stems branched, rooting almost their whole length, thickly covered with tiny leaves about 3 mm long; leaves each tipped by a minute bristle; bristles forming conspicuous tufts at the tips of the branches; fertile spike 10–25 mm long; sporophylls overlapping, more or less triangular. *Selaginella densa* Rydb. prairie selaginella. Dry grassland escarpment; localized.

3. EQUISETACEAE   horsetail family

Equisetum   **horsetail**

1a.   Stems bearing numerous branches in whorls at the nodes .... (2)
1b.   Stems unbranched .... (6)

2a.   Fertile and sterile stems similar, green; first internode of primary branches equaling or mostly shorter than the stem sheath; producing cones in summer .... (3)
2b.   Fertile and sterile stems not alike; first internode of the primary branches considerably longer than the stem sheath; producing cones in spring .... (4)

Fig. 1. *Lycopodium annotinum*, 1/2×.

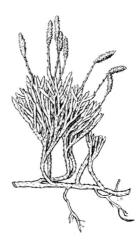

Fig. 3. *Lycopodium complanatum*, 2/5×.

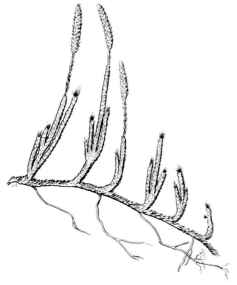

Fig. 2. *Lycopodium clavatum* var. *monostachyon*, 2/5×.

21

3a. Central cavity about four-fifths the diameter of the stem; stems to 1 m long, annual, single, but often forming dense stands, unbranched, or branches occurring sporadically or verticillate; teeth narrowly pointed. **Equisetum fluviatile** L.; water horsetail; Fig. 5. In water, bordering ponds and lakes and wet seepage slopes; occasional.

3b. Central cavity about one-sixth the diameter of the stem; stems annual, 20–80 cm long, solitary or clustered; branches spreading, in regular whorls from the middle nodes or few to none; teeth long, narrow, black with scarious margins. **Equisetum palustre** L.; marsh horsetail; Fig. 6. Wet lakeshores and ditches; occasional.

4a. Stem sheath teeth chestnut brown, papery; branches usually branched again; stems annual, of two kinds, erect, mostly solitary; fertile stem unbranched and lacking chlorophyll, becoming green and branched after the spores are released. **Equisetum sylvaticum** L.; wood horsetail; Fig. 7. Moist wooded areas and clearings; occasional.

4b. Stem sheath teeth dark, stiff; branches not branched again .... (5)

5a. Branches ascending; teeth of branch sheaths lance-attenuate; stems of two kinds, annual; sterile stems upright to prostrate or diffusely branched; sheaths with 4–14 teeth; teeth short, narrow, dark, scarious-margined, occasionally cohering in pairs; fertile stems lacking chlorophyll, precocious, withering and dying after spores are shed. **Equisetum arvense** L.; field horsetail; Fig. 8. Damp open woods, low open ground, roadside fill and embankments; frequent and often weedy.

5b. Branches horizontal to spreading; teeth of branch sheaths deltoid; stems of two kinds, annual, mostly solitary; sterile stems whitish green; sheaths pale; teeth narrow, persistent, white-margined, and dark-centered; fertile stems at first unbranched and lacking chlorophyll, precocious, becoming green and branched after the spores are shed. **Equisetum pratense** Ehrh.; meadow horsetail; Fig. 9. Moist open woodland; occasional to rare.

Fig. 4. *Selaginella selaginoides*, 1/2×.

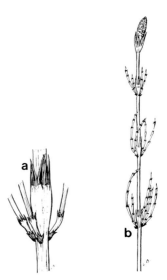

Fig. 6. *Equisetum palustre*, *a*, 2×; *b*, 1/2×.

Fig. 5. *Equisetum fluviatile*, 1/4×.

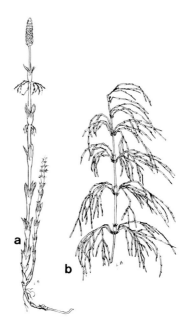

Fig. 7. *Equisetum sylvaticum* var. *pauciramosum*, *a*, 2/5×; *b*, 1/2×.

23

6a. Stems lacking chlorophyll, terminating in a spore-bearing cone. See *Equisetum arvense* and *E. pratense* (fertile stems).

6b Stems green . . . . (7)

7a. Stems annual; central cavity four-fifths the diameter of the stem; cones not sharp-tipped. See *E. fluviatile.*

7b. Stems perennial; cones sharp-tipped . . . . (8)

8a. Stems lacking a central cavity, caespitose, ascending or prostrate, arched-recurving, flexuous; sheaths green below, black above, loose, with 3 or rarely 4 deltoid scarious-margined teeth. **Equisetum scirpoides** Michx; dwarf scouring-rush; Fig. 10. In wet moss in coniferous woodland, and in open calcareous fen; rare.

8b. Stems with a central cavity . . . . (9)

9a. Stems with the central cavity one-third to two-thirds the diameter, tufted, ascending; sheaths green at the base, black above, slightly spreading; teeth 4–10, persistent, lanceolate to lance-deltoid, white-margined. **Equisetum variegatum** Schleich.; variegated horsetail; Fig. 11. In *Sphagnum* in Cold Spring Bog; apparently localized.

9b. Stems with central cavity three-quarters or more the diameter, upright, single or several together; sheaths developing dark bands at the base and summit, with the part between white or ashy gray; teeth numerous, lanceolate, promptly deciduous. **Equisetum hyemale** L. ssp. **affine** (Engelm.) Stone, scouring-rush; Fig. 12. Sandy and gravelly river and lake terraces; localized.

4. OPHIOGLOSSACEAE   adder's-tongue family

Botrychium   **grape fern**

1a. Small plants; sterile blades narrowly oblong, inserted below the middle, pinnate; segments opposite, obovate, rhomboidal or oblong; fertile segment narrowly paniculate. **Botrychium minganense** Vict. (*B. lunaria* (L.) Swartz var. *minganense* (Vict.) Dole). Meadow; apparently rare, although perhaps overlooked.

1b. Sterile blades triangular, wider than long . . . . . . (2)

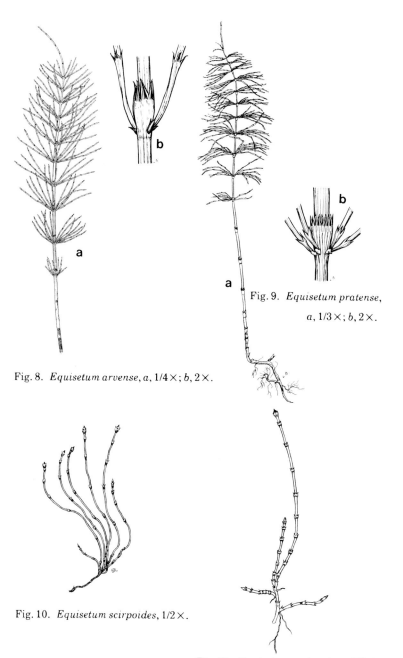

Fig. 8. *Equisetum arvense*, a, 1/4×; b, 2×.

Fig. 9. *Equisetum pratense*,
a, 1/3×; b, 2×.

Fig. 10. *Equisetum scirpoides*, 1/2×.

Fig. 11. *Equisetum variegatum*, 1/2×.

25

2a.　Blades very fleshy, evergreen, long-petioled, ternate, attached near the base of the plant; ultimate divisions crowded, sometimes imbricate, obtuse, or acutish; fertile segment usually broadly paniculate. *Botrychium multifidum* (Gmel.) Rupr.; leathery grape fern. Prairie; rare.

2b.　Deciduous; blades broadly deltoid, sessile, attached above the middle of the plant; the ultimate segments oblong-lanceolate, toothed, membranous or slightly fleshy (or the blade leathery, with its segments less toothed and often crowded and overlapping in var. *europaeum* Angstr.); fertile segment pinnately compound. *Botrychium virginianum* (L.) Swartz; rattlesnake fern; Fig. 13. Rich moist woodland and clearings; occasional.

5. PTERIDACEAE   fern family

Pteridium   **bracken**

Coarse fronds to 70 cm long, often forming extensive colonies; stipes about as long as the blade; blade triangular, usually ternate, bipinnate-pinnatifid to tripinnate-pinnatifid; ultimate divisions oblong to linear; margins revolute, covering the sori. *Pteridium aquilinum* (L.) Kuhn. var. *latiusculum* (Desv.) Underw.; bracken. Clearings in light soils; localized.

6. ASPIDIACEAE   fern family

1a.　Sterile and fertile fronds markedly different . . . . *Matteuccia*

1b.　Sterile and fertile fronds similar . . . . (2)

2a.　Fronds more or less ternate; indusium lacking . . . . *Gymnocarpium*

2b.　Fronds pinnate; pinnae variously divided; indusium present . . . . (3)

3a.　Sori elongate, often curved over the ends of the veins . . . . *Athyrium*

3b.　Sori round . . . . (4)

4a.　Indusium attached by its base on the side toward the midrib, hood-shaped . . . . *Cystopteris*

Fig. 12. *Equisetum hyemale* ssp. *affine*, 1/2 ×.

Fig. 13. *Botrychium virginianum* ssp. *europaeum*, 1/4 ×.

27

4b.   Indusium roundish-reniform, attached by a stalk in the centre .... ***Dryopteris***

Athyrium   **lady fern**

Fronds up to 1 m long, tufted and erect, spreading from stout ascending rhizomes; blades narrowly to broadly lanceolate, bipinnate to tripinnate; pinnae lanceolate; pinnules somewhat lobed to deeply toothed. ***Athyrium filix-femina*** (L.) Roth var. ***michauxii*** (Spreng.) Farw.; lady fern. Moist wooded areas; occasional.

Cystopteris   **bladder fern**

Fronds up to 35 cm long or longer, tufted from short creeping rhizomes; blades lanceolate, bipinnate; pinnae pinnatifid to lobed. ***Cystopteris fragilis*** (L.) Bernh.; fragile fern; Fig. 14. Steep shaded shale slope near East Gate and in shade in deep feather moss near Moon Lake; rare.

Dryopteris   **wood fern**

1a.   Basal pinnules on basal pinnae stalked; fronds 30–80 cm long, forming a crown at the top of a stout ascending rhizome; blades lanceolate, bipinnate or bipinnate-pinnatifid; pinnules oblong, with spine-tipped teeth. ***Dryopteris carthusiana*** (Vill.) H.P. Fuchs (*D. spinulosa* (O.F. Muell.) Watt.); spinulose wood fern; Fig. 15. Moist wooded areas; frequent.

1b.   Basal pinnules on basal pinnae sessile or adnate; fronds 25–70 cm long, forming a crown at the top of the stout ascending rhizome; fertile fronds longer than the sterile; blades linear-oblong to narrowly lance-oblong, pinnate-pinnatifid; basal pinnae short triangular. ***Dryopteris cristata*** (L.) Gray (*Thelypteris cristata* (L.) Nieuwl.); crested wood fern. Moist deciduous woodland; rare.

Gymnocarpium   **oak fern**

Delicate fronds up to 30 cm long arising singly from a slender forking rhizome; blades ternate, with the three divisions pinnate-pinnatifid; pinnules oblong, blunt. ***Gymnocarpium dryopteris*** (L.) Newm. ssp. ***dryopteris*** (*Dryopteris disjuncta* Am. auth.); oak fern; Fig. 16. Rich moist woodland; occasional.

Fig. 14. *Cystopteris fragilis*, 1/2×.

Fig. 16. *Gymnocarpium dryopteris*, 1/5×.

Fig. 15. *Dryopteris carthusiana*, 1/5×.

Matteuccia   **ostrich fern**

Fronds dimorphic, forming a crown at the end of the stout, widely creeping and forking rhizome; sterile fronds up to 1 m long, 12–24 cm wide, abruptly narrowed to the base, pinnate-pinnatifid; pinnae broadly linear, acuminate; pinnules oblong, bluntish; fertile fronds much shorter, persistent over winter; sori borne on the margins of the shallowly lobed, tightly inrolled, pod-like pinnae. **Matteuccia struthiopteris** (L.) Tod. var. **pensylvanica** (Willd.) Mort.; ostrich fern; Fig. 17. In low moist, often shaded situations; occasional.

7. PINACEAE   pine family

In addition to the species treated below, scots pine, red pine, white pine, western white pine, Siberian larch, and Norway spruce have been set out in plantations, particularly north of Clear Lake. Not all have survived.

1a.	Shrubs; cones berry-like, blue	***Juniperus***
1b.	Trees; cones woody	(2)
2a.	Leaves small, flat, closely imbricated, persistent	***Thuja***
2b.	Leaves needle-like, persistent or deciduous	(3)
3a.	Leaves in fascicles	(4)
3b.	Leaves borne singly	(5)
4a.	Leaves 2 per fascicle, evergreen	***Pinus***
4b.	Leaves many per fascicle, deciduous	***Larix***
5a.	Leaves flat, blunt-tipped	***Abies***
5b.	Leaves quadrangular, sharp-tipped	***Picea***

Abies   **fir**

Tall trees; branches horizontal; leaves sessile, whitened on the lower surface, appearing 2-ranked; cones erect, maturing the first year; scales deciduous. **Abies balsamea** (L.) Mill.; balsam fir. Wooded slopes, usually with white spruce; occasional.

Fig. 17. *Matteuccia struthiopteris* var. *pensylvanica*, 1/8×.

Juniperus   juniper

1a.   Decumbent shrub, sometimes forming large mats; leaves straight, awl-shaped, sharp-pointed, with a white stripe above. ***Juniperus communis*** L. var. ***depressa*** Pursh; common juniper; Fig. 18. Clearings and in partial shade, usually on lighter soils; occasional.

1b.   Prostrate and sometimes creeping shrub; leaves scale-like, overlapping. ***Juniperus horizontalis*** Moench; creeping juniper; Fig. 19. Dry open banks and slopes; rare.

Larix   larch

    Trees up to 10 m or higher, the branches slightly ascending; leaves soft, light green, turning yellow and falling in the autumn, leaving short spur shoots that are very obvious in winter; cones small, about 1 cm long, erect, falling in the second season. *Larix laricina* (Du Roi) K. Koch; larch, tamarack; Fig. 20. Scattered in bogs and wet places.

Picea   spruce

1a.   Branchlets glabrous; cones nearly cylindrical, falling after seed is shed; tall symmetrical tree with spreading branches; leaves needle-like, sharp-pointed, scattered around the twigs on peg-like bases. *Picea glauca* (Moench) Voss; white spruce; Fig. 21. A common upland tree.

1b.   Branchlets pubescent; cones ovoid to subglobose, persistent; narrow symmetrical trees with spreading branches; leaves needle-like, usually shorter and somewhat blunter, blue green. *Picea mariana* (Mill.) BSP; black spruce; Fig. 22. Wet or boggy woodlands.

Pinus   pine

Straight or gnarled trees; leaves often twisted and spreading; cones about 5 cm long, usually in pairs, curved and pointing toward the tips of the branches, remaining on the tree for many years and opening to release seeds only in the heat of a fire. *Pinus banksiana* Lamb.; jack pine; Fig. 23. Forms dense stands, particularly on lighter soils in the eastern part of the Park. There are some plantings west of Highway 10.

Thuja   arborvitae

Small conical trees with scale-like appressed and overlapping leaves on the flattened fan-shaped sprays of twigs; cones woody, oblong, erect, about 1 cm long, persisting over winter. *Thuja occidentalis* L.; eastern white cedar. Planted around buildings.

8. TYPHACEAE   cattail family

Typha   cattail

Plants to 1 m high or higher forming dense stands from thick creeping rhizomes; leaves long, linear, clasping; flowers

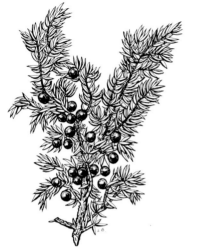

Fig. 18. *Juniperus communis*, 2/5×.

Fig. 20. *Larix laricina*, 3/5×.

Fig. 19. *Juniperus horizontalis*, 2/5×.

in a cylindrical spike, with the male above the female. **Typha latifolia** L.; common cattail; Fig. 24. Marshes, swamps, and ditches.

9. SPARGANIACEAE   bur-reed family

Sparganium   **bur-reed**

1a. Stigmas 2; fruits sessile, obpyramidal; plants robust terrestrial or emergent aquatic, to 150 cm high; leaves flat, up to 10 mm wide; inflorescence branched, with 1–3 pistillate heads and up to 20 staminate heads. **Sparganium eurycarpum** Engelm.; bur-reed; Fig. 25. Swamps and borders of lakes and streams; occasional.

1b. Stigmas 1; fruits stipitate, tapering at both ends; plants immersed, aquatic; upper leaves floating (or plants sometimes stranded on muddy shores); inflorescence usually unbranched . . . . (2)

2a. Leaves 2–5 mm wide, very elongate and usually floating, rounded on the back; fruiting heads 2–4, 1–2 cm in diameter; staminate heads 2–4. **Sparganium angustifolium** Michx.; bur-reed; Fig. 26. Shallow or deep water and sometimes on muddy shores.

2b. Leaves 5–10 mm wide, ribbon-like, flat on the back; fruiting heads 2–5, occurring above the axils, (1.5)2–2.5 cm in diameter; staminate heads 2–4, crowded. **Sparganium multipedunculatum** (Morong) Rydb.; bur-reed; Fig. 27. Marshes and borders of lakes and streams.

10. POTAMOGETONACEAE   pondweed family

Potamogeton   **pondweed**

1a. Submersed leaves 4 mm wide or wider . . . . (2)
1b. Submersed leaves less than 4 mm wide . . . . (6)

2a. Base of submersed leaves cordate or auriculate . . .(3)
2b. Base of submersed leaves neither cordate nor auriculate . . . . (4)

3a. Stipules usually persistent and conspicuous; leaves mostly 4–25 cm long, often boat-shaped at the uninjured tip; flowers in 6–12 whorls, forming a dense

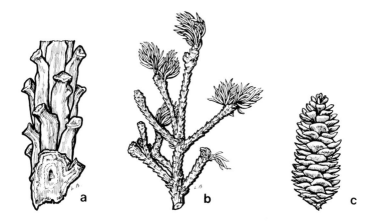

Fig. 21. *Picea glauca*, *a*, 5×; *b*, 1/2×; *c*, 1/2×.

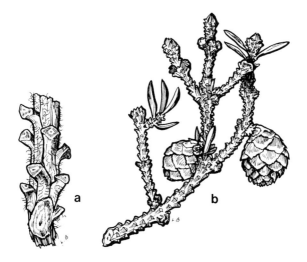

Fig. 22. *Picea mariana*, *a*, 4×; *b*, 4/5×.

35

spike in fruit. **Potamogeton praelongus** Wulf.; Fig. 28. Deep cold waters of lakes and streams.

3b. Stipules soon disintegrating into fibers and disappearing; leaves 1.5–10 cm long; flowers in 6–12 whorls, moniliform, but forming a dense spike in fruit. **Potamogeton richardsonii** (Ar. Benn.) Rydb.; Fig. 29. Lakes and rivers.

4a. Stems strongly wing-flattened, 1–3 mm wide; leaves linear, not more than 5 mm wide, with 1 or 3 strong nerves and many fine ones; stipules firm, 1.5–3.5 cm long; fruiting spikes cylindrical, with 7–11 whorls. **Potamogeton zosteriformis** Fern.; Fig. 30. Quiet waters.

4b. Stems nearly round in cross section or, if flattened, slender .... (5)

5a. Submersed leaves mostly 8–14 cm long, reddish brown to olive green, transparent; floating leaves (when present) thin and delicate; blades tapered to the base, not sharply distinct from the petiole; flowering spike 5–9 whorls, rather open but crowded in fruit. **Potamogeton alpinus** Balbis var. **tenuifolius** (Raf.) Ogden; Fig. 31. Lakes and rivers.

5b. Submersed leaves mostly 3–8 cm long, green; floating leaves firm; blades more or less rounded at the base, distinct from the petiole; spike compact, in 5–10 whorls. **Potamogeton gramineus** L.; Fig. 32. Very variable depending on depth of water; frequently stranded.

6a. Stipules sheathing the base of the leaf, with only their summits free .... (7)

6b. Stipules free to the base .... (8)

7a. Stipular sheaths (at least of the lower primary leaves) loose and much wider than the stem; spikes 3–8 cm long, in 5–12 nearly equidistant whorls. **Potamogeton vaginatus** Turcz.; Fig. 33. Quiet water.

7b. Stipular sheaths tight, scarcely wider than the stem; spikes 0.5–5 cm long, with the lower whorls progressively farther apart. **Potamogeton pectinatus** L.; Fig. 34. Quiet water.

8a. Submersed leaves reduced to the petioles only, firm, 10–20 cm long; floating leaves leathery in texture, shining, attached to the petiole by a brownish joint about 1.5 cm long; spike compact in 8–14 whorls, in

Fig. 23. *Pinus banksiana*, 1×.

Fig. 25. *Sparganium eurycarpum*, 1/3×.

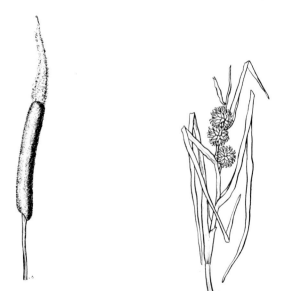

Fig. 24. *Typha latifolia*, 1/5×.   Fig. 26. *Sparganium angustifolium*, 2/5×.

fruit 3–5 cm long. **Potamogeton natans** L.; Fig. 35. Lakes and quiet streams.

8b.   Submersed leaves with flat blades .... (9)

9a.   Floating leaves usually well developed. See *Potamogeton gramineus.*

9b.   Floating leaves absent .... (10)

10a.   Stem strongly flattened. See *Potamogeton zosteriformis.*

10b.   Stem slender, nearly round in cross section; leaves 3-nerved, acute to obtuse, usually with a pair of translucent basal glands; peduncles upwardly broadened; spike cylindrical, interrupted. **Potamogeton strictifolius** Ar. Benn. var. **rutiloides** Fern. Quiet water of shallow ponds, streams, and lakeshores.

11. NAJADACEAE

Najas   naiad

Plants submersed, aquatic, with much-branched slender stems and narrow ribbon-like leaves; leaves enlarged at the base; flowers single, inconspicuous, borne in axils of the leaves; seeds shining. **Najas flexilis** (Willd.) Rostk. & Schmidt; naiad; Fig. 36. Rooted in muck in shallow water of slow-moving streams and borders of lakes and ponds.

12. SCHEUCHZERIACEAE   arrow-grass family

1a.   Flowers in a bracted few-flowered raceme; stems leafy .... **Scheuchzeria**

1b.   Flowers in a bractless many-flowered spikelike raceme; stems leafless .... **Triglochin**

Scheuchzeria

Leaves sheathing and ligulate like a grass; flowers small and inconspicuous; fruit consisting of 3 spreading follicles. **Scheuchzeria palustris** L.; Fig. 37. Quaking fen; apparently rare and localized.

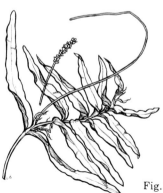

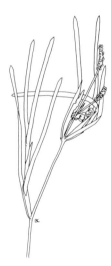

Fig. 29. *Potamogeton richardsonii*, 2/5×.

Fig. 27. *Sparganium multipedunculatum*, 2/5×.

Fig. 30. *Potamogeton zosteriformis*, 2/5×.

Fig. 28. *Potamogeton praelongus*, 1/4×.

Triglochin   **arrow-grass**

1a.   Sepals broadly rounded; fruit oblong; carpels 6; plants robust; leaves all basal, narrow, and elongate. ***Triglochin maritimum*** L.; arrow-grass; Fig. 38. Marshes and fens.

1b.   Sepals acuminate; fruit narrowly oblanceolate; carpels 3. Similar to the previous species but more delicate and with finer leaves. ***Triglochin palustre*** L.; slender arrow-grass; Fig. 39. Open, calcareous fen; rare.

13. ALISMATACEAE   water-plantain family

1a.   Leaves ovate; flowers perfect, small; petals about 5 mm long . . . . ***Alisma***

1b.   Leaves sagittate; flowers unisexual, larger; petals about 10–12 mm long . . . . ***Sagittaria***

Alisma   **water-plantain**

Leaves basal, long-petioled, ascending; inflorescence much branched, standing above the leaves; flowers perfect; fruit a ring of carpels. ***Alisma triviale*** Pursh (*A. plantago-aquatica* of auth.); water-plantain; Fig. 40. Ditches, sedge meadows, and marshes, rooted in mud.

Sagittaria   **arrowhead**

1a.   Achenes 2.0–2.6 mm long; beak slender, erect or incurved, borne from the summit well in from the margin; leaves basal, long-petioled, mostly sagittate or if in deep water, ribbon-like; inflorescence in 2–5 whorls; flowers monecious; carpels in dense heads. ***Sagittaria cuneata*** Sheld; arrowhead; Fig. 41. Muddy stream banks and swamps; apparently rare.

1b.   Achenes 2.3–3.5 mm long; beak marginal, broad-based; plant otherwise similar to *S. cuneata*. ***Sagittaria latifolia*** Willd.; arrowhead. Muddy lakeshores, stream banks, marshes, and ditches; occasional, more frequent than *S. cuneata*.

Fig. 31. *Potamogeton alpinus* ssp. *tenuifolius*, 1/3×.

Fig. 33. *Potamogeton vaginatus*, 1 2/3×.

Fig. 34. *Potamogeton pectinatus*, 3/5×.

Fig. 32. *Potamogeton gramineus*, 1/8×.

41

14. HYDROCHARITACEAE   frog's-bit family

Elodea   **waterweed**

Submersed plants with branching stems; leaves oblong-ovate, about 6 mm long, in whorls of 2–4; flowers rarely found; pistillate flowers with a long thread-like tube that reaches the surface. *Elodea canadensis* Michx. (*Anacharis canadensis* (Michx.) Planch.). Forms thick stands rooted in muck in quiet waters.

15. GRAMINEAE   grass family   Fig. 42

1a.	Spikelets gathered in 1 or more spikes . . . . (2)
1b.	Inflorescence a panicle of spikelets that is sometimes narrow and spiciform or is rarely reduced to a raceme or to 1 or a few spikelets . . . . (11)
2a.	Inflorescence a single spike . . . . (3)
2b.	Inflorescence of 2 or more spikes . . . . (7)
3a.	Spikelets 2 or 3 to each node of the rachis . . . . (4)
3b.	Spikelets 1 to a node . . . . (5)
4a.	Spikelets 3 to a node . . . . **Hordeum**
4b.	Spikelets 2 to a node . . . . **Elymus**
5a.	Spikelets with only 1 glume (the outer one); spikelet borne edgewise to the rachis . . . . **Lolium**
5b.	Both glumes present; spikelet borne sidewise to the rachis . . . . (6)
6a.	Glumes and lemmas strongly asymmetrical, tending to be 3-toothed at the tip . . . . **Triticum**
6b.	Glumes and lemmas entire at the tip and quite symmetrical; keel located along the middle . . . . **Agropyron**
7a.	Spikes borne together in a terminal, digitate or subdigitate cluster . . . . **Andropogon**
7b.	Spikes in a raceme or panicle . . . . (8)
8a.	Inflorescence a raceme of 2 or more spikes . . **Spartina**
8b.	Inflorescence a panicle, open or narrow and spiciform . . . . (9)

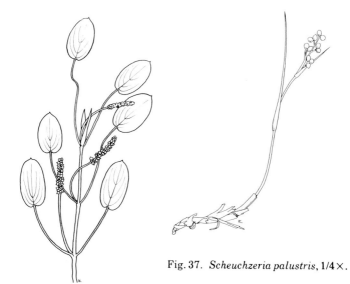

Fig. 37. *Scheuchzeria palustris*, 1/4×.

Fig. 35. *Potamogeton natans*, 1/5×.

Fig. 36. *Najas flexilis*, 2/5×.

Fig. 38. *Triglochin maritimum*, 1/2×.

9a. Spikes lax, in 2 vertical ranks . . . . **Andropogon**
9b. Spikes strongly secund . . . . (10)

10a. Glumes laterally compressed and enclosing the spikelet . . . . **Beckmannia**
10b. Glumes somewhat dorsally flattened and less than half as long as the spikelet . . . . **Echinochloa**

11a. Panicle spiciform . . . . (12)
11b. Panicle more obvious, often open and lax or with the lower branches longer than the spikelets . . . . (16)

12a. Spikelets awnless . . . . (13)
12b. Spikelets with awns . . . . (14)

13a. Spike ovoid, over 1 cm wide . . . . **Phalaris**
13b. Spike linear and much narrower . . . . **Alopecurus**

14a. Awns subtending the spikelet . . . . **Setaria**
14b. Awns terminating the glumes or lemmas . . . . (15)

15a. Glumes awned; lemmas awnless, acicular-ciliate on the keel . . . . **Phleum**
15b. Glumes awnless; lemmas with an obvious dorsal awn . . . . **Alopecurus**

16a. Functional florets 2 or more per spikelet . . . . (17)
16b. Only 1 functional floret per spikelet . . . . (37)

17a. Glumes overtopping the spikelets or at least the upper glume reaching to the summit of the lowest lemma (awns excluded) . . . . (18)
17b. Glumes shorter, the lowest lemma at least as long as the upper glume and overtopping it . . . . (27)

18a Spikelets awnless or with awns arising from the tip of the lemmas . . . . (19)
18b. Lemmas awned, with the awn arising below the tip of the lemma . . . . (23)

19a. Spikelet suborbicular . . . . **Hierochloe**
19b. Spikelet elongate, more or less lanceolate . . . . . . (20)

20a. Spikelets few, each primary branch with only 1 or 2 (–3) spikelets . . . . **Festuca**
20b. Spikelets more numerous . . . . (21)

Fig. 39. *Triglochin palustre*, 3/5×.

Fig. 41. *Sagittaria cuneata*, 1/4×.

Fig. 40. *Alisma triviale*, 1/4×.

45

21a. Panicle cylindrical; branches uniformly short and less than 1 cm long .... **Koeleria**

21b. Branches of panicle much longer, with the lower ones 2–4 times longer than the upper ones .... (22)

22a. Callus glabrous .... **Festuca**

22b. Callus with a tuft of hairs 0.5 mm long or longer .... **Scolochloa**

23a. Lemma bifid at tip; awn arising from the bottom of the sinus .... **Danthonia**

23b. Awn arising from the back of the lemma .... (24)

24a. Lemma bidentate at the tip; spikelet 1 cm long or longer (excluding the awns) .... (25)

24b. Lemma merely erose at the tip; spikelet much shorter .... (26)

25a. Branches bearing only 1 or 2 spikelets .... **Helictotrichon**

25b. Longer branches bearing many spikelets ..... **Avena**

26a. Spikelet 3-flowered .... **Hierochloe**

26b. Spikelet 2-flowered .... **Deschampsia**

27a. Rachilla long-bearded; hairs overtopping the florets .... **Phragmites**

27b. Rachilla not bearded or the hairs much shorter .... (28)

28a. Spikelets short, not much longer than the glumes; lowest lemma about equaling the tip of the upper glume .... (29)

28b. Spikelets more elongate; lowest lemma much overtopping the upper glume .... (30)

29a. Upper glume at least twice as large as the lower; spikelet disarticulating below the glumes .... **Sphenopholis**

29b. Glumes nearly similar; spikelet disarticulating above the glumes .... **Koeleria**

30a. Florets successively smaller, with the 1(-3) upper florets reduced to a much smaller and sterile lemma; callus bearded .... **Schizachne**

30b. All florets similar or the upper slightly reduced .... (31)

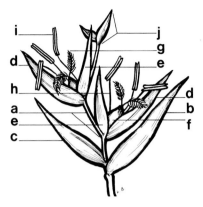

a. rachilla
b. first glume
c. second glume
d. lemma
e. palea
f. ovary
g. stigma
h. filament
i. anther
j. sterile floret

Fig. 42. Stylized grass spikelet, 1/2 ×.

31a.   Lemma minutely to obviously bifid at the tip, often aristate and nearly always over 1 cm long ... **Bromus**
31b.   Lemma entire at the tip, mostly not aristate and mostly shorter .... (32)

32a.   Lemmas keeled .... **Poa**
32b.   Lemmas rounded on the back .... (33)

33a.   Lemma subulate to aristate at the tip .... **Festuca**
33b.   Lemma obtuse to rounded and narrowly to broadly membranous-margined at the tip .... (34)

34a.   Lemma with 5–9 obvious and about equally raised nerves .... (35)
34b.   Lemma almost nerveless or with the midnerve conspicuous .... (36)

35a.   Leaf sheath forming a closed cylinder, its margins fused ventrally .... **Glyceria**
35b.   Sheath margins free .... **Torreyochloa**

36a.   Nerves of lemma about equally faint ..... **Puccinellia**
36b.   Midnerve much stronger and clearly raised ..... **Poa**

37a.   Lemma (and glumes) awnless .... (38)
37b.   Lemma (or glumes) bearing a dorsal or terminal awn .... (44)

38a.   Glumes strongly differentiated .... (39)
38b.   Glumes similar to the lemma .... (40)

39a. Spikelets and lemmas broadly obovoid to suborbicular
.... ***Echinochloa***
39b. Spikelets and lemmas much longer than wide ......
.... ***Sporobolus***

40a. Lemma coriaceous, of much harder texture than the
glumes .... (41)
40b. Lemma similar to the glumes or of thinner texture
.... (42)

41a. Panicle crowded, cylindric .... ***Phalaris***
41b. Panicle open; its branches spreading more or less
horizontally .... ***Milium***

42a. Palea obscure or very much smaller than the lemma
.... ***Agrostis***
42b. Palea similar to the lemma and about as long ... (43)

43a. Lemma 1-nerved .... ***Sporobolus***
43b. Lemma with faint lateral nerves ..... ***Muhlenbergia***

44a. Spikelet subtended by a suborbicular glume, less than
half as long as the spikelet .... ***Echinochloa***
44b. Glumes nearly as long as, to much longer than, the
floret (awns excluded) .... (45)

45a. Awns over 1 cm long, geniculate and twisted .... (46)
45b. Awns shorter .... (47)

46a. Awns 2 cm long or more, persistent .... ***Stipa***
46b. Awns shorter, more or less deciduous ..... ***Oryzopsis***

47a. Awn arising dorsally on the lemma .... (48)
47b. Awn terminal .... (49)

48a. Callus not barbellate .... ***Agrostis***
48b. Callus bearing a tuft of hairs often as long as the
spikelet .... ***Calamagrostis***

49a. Lemma of a much harder structure than the glumes
and tightly enclosing the seed .... ***Oryzopsis***
49b. Lemma and glumes of a similar structure or the
lemma thinner than the glumes .... (50)

50a. Lemma bidentate; short awn arising from between the
teeth .... ***Cinna***
50b. Lemma and glumes entire; awned or awnless ......
.... ***Muhlenbergia***

Agropyron — wheat grass

1a. Plants with rhizomes . . . . (2)
1b. Plants with fibrous roots . . . . (3)

2a. Leaves (or at least the larger ones) 5–10 mm wide; culms up to 100 cm long, arising in tufts from the long, thick, yellowish white rhizomes; stomata visible as white lines on the underside of the leaves; spikes up to 15 cm long; spikelets 4–7 flowered; glumes usually smooth, awn-tipped; anthers (3-) 4–6 (-7) mm long. *Agropyron repens* (L.) Beauv.; quack grass, couch grass. Introduced roadside and garden weed.

2b. Leaves 1–4 mm wide; culms up to 60 cm long, arising in tufts along the creeping rhizomes; leaves stiffly divergent, glaucous; spikes 7–15 cm long; spikelets 6–10 flowered; glumes tapering to a short sharp awn; anthers 3.5–4.5 mm. *Agropyron smithii* Rydb.; western wheat grass. Prairie; occasional.

3a. Spikes 2–7 cm long, flat; spikelets very closely spaced on the rachis; culms up 30–50 cm, densely tufted. *Agropyron cristatum* (L.) Gaertn.; crested wheat grass. Introduced forage species; occasionally escaped.

3b. Spikes elongated, 6–25 cm long, not flat; spikelets not closely spaced on the rachis; culms up to 100 cm long, loosely tufted; four varieties in our area: var. *trachycaulum*, forming large leafy and loose tussocks; culms to 1 m; spikes slender; lowermost spikelets usually somewhat remote; var. *novae-angliae* (Scribn.) Fern., similar, but culms shorter and commonly fewer together; spikes shorter and denser; spikelets overlapping and commonly in two distinct rows; var. *glaucum* (Pease & Moore) Malte; plant glaucus throughout; awn of the lemma very thin and usually as long as the body; and var. *unilaterale* (Cassidy) Malte (*A. subsecundum* (Link.) Hitchc.), forming small, loose and leafy tussocks; spikes dense, erect, slightly curved or nodding and slightly one-sided; awns of the lemma thicker and longer. *Agropyron trachycaulum* (Link) Malte; Fig. 43. Slender wheat grass. Open woodland, prairie patches, clearings, and disturbed situations. Hybridizes with *Hordeum jubatum* to form ×*Agrohordeum macounii* (Vasey) Lepage.

Agrostis   bent grass

1a.   Plants tufted, delicate; culms erect, up to 70 cm long; panicle very diffuse; spikelets up to 2.7 mm long. ***Agrostis scabra*** Willd.; hair grass; tickle grass; Fig. 44. Open woods, clearings, and waste places; frequent.

1b.   Plants robust, often rhizomatous, sometimes rooting at the nodes of decumbent culms, up to 80 cm high; panicle becoming open; spikelets mostly longer. ***Agrostis stolonifera*** L. (*A. alba* L.); redtop. Cultivated and escaping to disturbed and wet situations.

Alopecurus   foxtail

Densely tufted; culms up to 50 cm long; inflorescence simulating that of timothy (*Phleum pratense*), but more delicate. ***Alopecurus aequalis*** Sobol.; foxtail; Fig. 45. Wet depressions, by streams and lakeshores; localized.

Andropogon   beard grass

Culms conspicuously purplish, up to 150 cm long, occurring in large tufts; rhizomes short; leaf blades blue green to glaucous; racemes 3–6, 5–10 cm long, usually purplish. ***Andropogon gerardii*** Vitman; bluestem. South-facing slopes; rare.

Avena   oats

1a.   Inflorescence open; spikelets drooping, mostly 3-flowered, disarticulating early; lemma pubescent; awns stout, 1–5 cm long, geniculate, blackish below the bend, pale green above; culms 30–70(–100) cm long; blades 4–8 mm wide. ***Avena fatua*** L.; wild oats. A noxious weed of grain fields and waste places.

1b.   Inflorescence secund; spikelets 2-flowered, not disarticulating readily at maturity; lemmas glabrous; awns small and straight or lacking; culms usually shorter than *A. fatua*. ***Avena sativa*** L.; oats. Waste ground by roadsides; spontaneous.

Beckmannia

Tufted pale green annual; culms up to 70 cm long; inflorescence a panicle or raceme up to 25 cm long, with its

Fig. 45. *Alopecurus aequalis*, 1×.

Fig. 43. *Agropyron trachycaulum*, 1/2×.

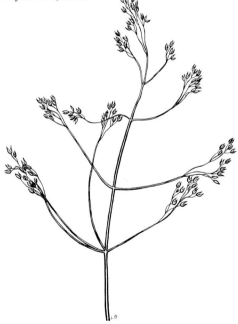

Fig. 44. *Agrostis scabra*, 3/4×.

branches appressed to ascending; branches bearing one-sided racemes of closely imbricated 1-flowered spikelets. **Beckmannia syzigachne** (Steud.) Fern.; slough grass; Fig. 46. Borders of ponds, streams, sloughs, and ditches; frequent.

Bromus   brome

1a. Plants with rhizomes . . . . (2)
1b. Plants with fibrous roots . . . . (3)

2a. Lemmas glabrous; blades and culms glabrous or somewhat scabrous; culms up to 100 cm long; spikelets usually purplish. **Bromus inermis** Leyss.; brome. An introduced forage grass; roadsides and waste places.
2b. Lemmas pubescent, at least along the margins; blades pilose above; culms pubescent at the nodes, up to 100 cm long. **Bromus pumpellianus** Scribn.; Fig. 47. Lake bank; rare.

3a. First glume 3-nerved; culms up to 60 cm long, slender, pubescent at the nodes; panicle nodding; branches flexuous, spreading or drooping; lemmas densely and evenly pubescent on the back. **Bromus porteri** (Coult.) Nash. Clearings and scrub prairie; apparently rare.
3b. First glume 1-nerved . . . . (4)

4a. Lemmas pubescent along the margins and lower part only, with the upper part glabrous; culms up to 100 cm long; sheaths glabrous or short pubescent at the nodes; panicle with the slender branches often spreading or drooping. **Bromus ciliatus** L.; fringed brome. Open woodland, clearings, and scrub prairie; frequent.
4b. Lemmas rather evenly pubescent on the back, but more densely so on the lower margins; culms up to 100 cm long, pubescent at the nodes; sheaths usually retrorsely pubescent; panicle nodding; branches elongate, spreading or drooping. **Bromus latiglumis** (Shear) Hitchc. (*B. purgans* L.); Canada brome. Clearings and scrub prairie; occasional.

Calamagrostis   reed grass

1a. Panicle open; branches spreading and often drooping; plants tufted; rhizomes creeping; culms up to 120 cm long or longer; nodes 5 or 6; blades lax; callus hairs about as long as the lemma; awn delicate, straight,

Fig. 46. *Beckmannia syzigachne*, 2/3×.

Fig. 47. *Bromus pumpellianus* var. *pumpellianus*, 4/5×.

53

inserted just below the middle. **Calamagrostis canadensis** (Michx.) Beauv.; blue-joint, marsh reed grass; Fig. 48. Lakeshores, borders of marshes, ditches, and moist open woodland; frequent.

1b. Panicle narrow and more or less contracted; branches appressed and ascending .... (2)

2a. Blades and upper part of culm scabrous; culms stiff,up to 100 cm long; leaves becoming involute; panicle pale, usually dense and spike-like; glumes firm and opaque, acute; callus hairs shorter than the lemma. **Calamagrostis inexpansa** A. Gray; Fig. 49. Clearings and borders of woods.

2b. Blades smooth or scabrous only at the tip; culms smooth except under the panicle; panicle stiff, usually brownish; glumes thin, hyaline, and somewhat translucent; callus hairs unequal, shorter than the lemma. **Calamagrostis neglecta** (Ehrh.) Gaertn., Mey. & Schreb.; Fig. 50. Wet meadows, ditches, and moist disturbed situations; frequent.

Cinna   wood grass

Tall, tufted grass; culms up to 150 cm high; leaves short, broad; panicle large, open, greenish or yellowish; branches slender, spreading or drooping. **Cinna latifolia** (Trev.) Griseb.; wood grass; Fig. 51. Moist open woods, thickets, and clearings; frequent.

Danthonia   oat grass

1a. Glumes glabrous, (1.2–)1.5(–1.7) cm long, purplish; lemma 5.5–8.0 mm long; teeth 1.5–2.0 mm long; awns with the upper segment purplish and the lower greenish or yellow; tufted grass up to 60 cm high or higher, with a closed and secund inflorescence of 3–10 spikelets; old leaves marcescent and curly. **Danthonia intermedia** Vasey; timber oat grass; Fig. 52. Prairie; occasional.

1b. Glumes usually pilose or strigose, (0.8–)1.0(–1.2) cm long; lemma 4–5 mm long, including the subulate teeth; awns deep brown or purple in the twisted portion; plant tufted as in *D. intermedia*. **Danthonia spicata** (L.) Beauv.; poverty oat grass; Fig. 53. Scrub prairie; less frequent than *D. intermedia*.

Fig. 48. *Calamagrostis canadensis*, 1/4×.

Fig. 50. *Calamagrostis neglecta*, 1/4×.

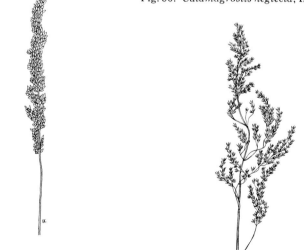

Fig. 49. *Calamagrostis inexpansa*, 2/5×.

Fig. 51. *Cinna latifolia*, 1/2×.

Deschampsia   hair grass

Densely tufted; culms up to 100 cm long; leaves narrow, often folded, mostly basal; panicle open and diffuse, more or less pyramidal; branches slender, more or less scabrous, bearing mostly 2-flowered largely hyaline spikelets towards the tips. **Deschampsia caespitosa** (L.) Beauv.; tufted hair grass; Fig. 54. Moist situations and fens; occasional.

Echinochloa   barnyard grass

Coarse tufted annual; culms erect to decumbent, up to 100 cm long; leaves flat or V-shaped; spikelets very irregularly awned, borne in a raceme or in a panicle of spiciform racemes; inflorescence often maturing to purplish black. **Echinochloa wiegandii** (Fassett) McNeill & Dore (*E. pungens* (Poir.) Rydb. var. *wiegandii* Fassett); barnyard grass. Weedy native species; occasional in disturbed situations.

Elymus   wild rye, lyme grass

1a.   Awns at least as long as the lemmas . . . . (2)
1b.   Awns less than half as long as the lemmas . . . . (4)

2a.   Awns straight; plant forming small, loose tufts; culms up to 100 cm long or longer; leaf blades flat, scabrous on both sides; spike up to 15 cm long, erect or somewhat flexuous. **Elymus virginicus** L.; Virginia wild rye. Grassy clearings; rare.
2b.   Awns long, arching outward . . . . (3)

3a.   Glumes about 1 mm wide, with 3–5 rugose nerves; culms up to 125 cm long or longer, tufted, with short rhizomes at least when young; leaf blades wide, flat or sometimes convolute, dark green to glaucous, smooth to scabrous on both sides; spike up to 25 cm long, nodding. **Elymus canadensis** L.; Canada wild rye. River and lake banks and clearings; occasional.
3b.   Glumes about 0.5 mm wide, 1-nerved or nerveless; culms up to 100 cm long, tufted; leaves wide, somewhat villous above, smooth below; spike to 15 cm

Fig. 52. *Danthonia intermedia*, 4/5×.

Fig. 54. *Deschampsia caespitosa*, 1/4×.

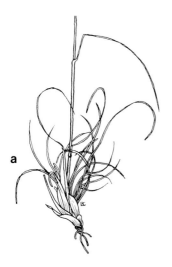

a

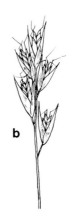

b

Fig. 53. *Danthonia spicata*, *a*, 2/5×; *b*, 4/5×.

long, flexuous or somewhat nodding. **Elymus diversiglumis** Scribn. & Ball (*E. interruptus* Buckl.). Open woodland and clearings; occasional.

4a.   Glumes less than 0.5 mm wide and more or less setaceous; culms up to 80 cm long, in small tufts from long creeping rhizomes; leaves flat to convolute, scabrous on both sides; spikes dense, up to 12 cm long, purplish; both glumes and lemmas villous. **Elymus innovatus** Beal; hairy wild rye; Fig. 55. Clearings, open woods, usually in lighter soils; frequent.

4b.   Glumes wider, flat. See *Elymus virginicus*.

Festuca   fescue

1a.   Leaf blades generally more than 3 mm wide, soft, flat, convolute in the shoot and inrolling on drying; plant stoloniferous and sod-forming; culms up to 100 cm long; panicle up to 20 cm long, contracted before and after flowering, often somewhat secund; spikelets linear-cylindrical, about twice as long as the upper glume; lemmas thin, awnless, with scarious margins. **Festuca pratensis** Huds. (*F. elatior* L. pro parte); meadow fescue. Introduced forage grass; occasional.

1b.   Leaf blades less than 3 mm wide, usually permanently folded, conduplicate in the shoot . . . . (2)

2a.   Glumes as long or almost as long as the spikelet; culms up to 100 cm long or longer, tufted, forming large tussocks; panicle up to 20 cm long, open to somewhat contracted; lemmas uniformly scabrous, sometimes with a very short awn. **Festuca hallii** (Vasey) Piper (*F. scabrella* Torr. pro parte); rough fescue. Open woodland and scrub prairie; abundant in native prairies.

2b.   Glumes distinctly shorter than the spikelet . . . . . . (3)

3a.   Plant rhizomatous, turf-forming; culms up to 80 cm long, erect to somewhat decumbent at the base; leaf sheaths reddish to purplish at the base; panicle up to 20 cm long, erect or somewhat nodding; lemmas often somewhat pubescent, awn-tipped. **Festuca rubra** L.; red fescue; Fig. 56. Introduced pasture and lawn grass; major plantings along roadsides.

3b.   Plant densely tufted; culms up to 30 cm long; leaves filiform, tightly rolled; panicle up to 7 cm long, narrow, contracted both before and after flowering; spikelets sometimes violet-tinged; lemmas awned.

Fig. 55. *Elymus innovatus*, 2/5×.

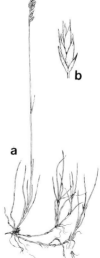

Fig. 56. *Festuca rubra*, *a*, 1/4×; *b*, 3×.

Festuca saximontana Rydb. (*F. ovina* L. var. *saximontana* (Rydb.) Gl.); Rocky Mountain fescue; Fig. 57. Scrub prairie, clearings, and disturbed situations; occasional.

Glyceria

manna grass

1a. Inflorescence closed, long-linear; spikelets 1 cm long or longer, linear-cylindrical; culms up to 100 cm long, solitary or in tufts from creeping rhizomes; leaf blades flat or folded; lemmas strongly 7-nerved. *Glyceria borealis* (Nash) Batch.; northern manna grass. Sloughs, lakeshores, and stream margins; occasional.

1b. Inflorescence open, broadly lanceolate to pyramidal; spikelets shorter and broader . . . . (2)

2a. Upper glume about 1 mm long, twice as long as the lower; culms up to 80 cm long, often in large clumps, from long, creeping rhizomes; panicle to 20 cm long, erect or nodding at the tip; lemma 7-nerved, more or less scarious at the tip. *Glyceria striata* (Lam.) Hitchc.; fowl manna grass; Fig. 58. Shallow water of marshes, mud bordering lakes and streams, and in ditches; frequent.

2b. Upper glume 2–2.5 mm long, not much exceeding the lower; glumes acute, whitish; lemmas purple, barely scarious-margined, strongly 7-nerved; culms up to 200 cm long, solitary or in tufts, from creeping rhizomes; panicle up to 40 cm long, rather dense, usually nodding. *Glyceria grandis* S. Wats.; tall manna grass; Fig. 59. Wet ground and shallow water of marshes and borders of lakes and streams; frequent.

Helictotrichon

Densely tufted; culms up to 40 cm long or longer; leaves flat or folded; margin and midnerve finely outlined in white; inflorescence narrow; branches erect or ascending; spikelets about 1.5 cm long, with 4 or 5 florets; glumes thin and greenish, somewhat shiny, about as long as the spikelet; lemmas 10–12 mm long; awn geniculate, up to 2 cm long, darker and twisted below the knee. *Helictotrichon hookeri* (Scribn.) Henr. (*Avena hookeri* Scribn.); spike-oat, oat grass; Fig. 60. Scrub prairie and parkland; occasional.

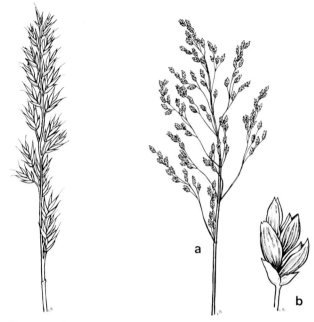

Fig. 57. *Festuca saximontana*, 1 1/5×.

Fig. 58. *Glyceria striata*, *a*, 3/5×; *b*, 4×.

Fig. 59. *Glyceria grandis*, 1/5×.

Fig. 60. *Helictotrichon hookeri*, 1/4×.

61

Hierochloe   holy grass

Culms up to 60 cm long, solitary or with a few leafy shoots, from a long creeping rhizome; panicle up to 15 cm long, pyramidal; spikelets lustrous golden yellow; sweet scented. ***Hierochloe odorata*** (L.) Beauv.; sweet grass; holy grass; Fig. 61. Borders of open woods, clearings, and scrub prairie; occasional.

Hordeum   barley

Culms densely tufted, up to 60 cm long; spike 5–10 cm long, often nodding; awns (3–)4–5(–7) cm long, very fine, rather uniform in length. ***Hordeum jubatum*** L.; foxtail barley, squirrel-tail grass; Fig. 62. Clearings, banks, and disturbed situations; common. Hybridizes with *Agropyron trachycaulum* to form ×*Agrohordeum macounii* (Vasey) Lepage.

Koeleria   June grass

Culms up to 50 cm long, densely tufted; basal leaves short to half the length of the culm, narrow; inflorescence a dense, cylindrical spike-like panicle up to 15 cm long; lemmas lustrous. ***Koeleria macrantha*** (Led.) Schultes (*K. cristata* (L.) Pers.); June grass; Fig. 63. Scrub prairie and in disturbed situations; frequent.

Lolium   rye grass, darnel

Perennial turf-forming species; culms up to 60 cm long; var. ***perenne*** has lemmas awnless or the awn less than 1 mm long; var. ***aristatum*** Willd. (*L. multiflorum* Lam.) is a more robust annual, biennial, or short-lived perennial, with at least some lemmas with an awn more than 1 mm long. ***Lolium perenne*** L.; rye grass. Introduced forage species; rare in waste places.

Milium   millet grass

Erect culms from a bent base, up to 70 cm long, from short stout rhizomes; leaf blades up to 12 mm wide, flat; panicles up to 20 cm long, open, pyramidal, with the slender branches spreading. ***Milium effusum*** var. ***cistatlanticum*** Fern.; millet grass. Understory in open woodland in the eastern parts of the Park; rare.

Fig. 61. *Hierochloe odorata*, 1/4×.

Fig. 62. *Hordeum jubatum*, 2/5×.

Fig. 63. *Koeleria macrantha*, 1/4×.

Muhlenbergia   muhly

1a. Panicle very narrowly linear, usually not more than 2 mm wide; leaf blades 1–2 mm wide . . . . (2)

1b. Panicle not narrowly linear, usually about 5 mm wide; leaf blades 2–8 mm wide . . . . (3)

2a. Glumes ovate, 1–1.5 mm long, less than half as long as the spikelets; culms up to 40 cm long, densely tufted from hard scaly rhizomes; inflorescence 3–10 cm long. *Muhlenbergia richardsonis* (Trin.) Rydb.; mat muhly; Fig. 64. Prairies; frequent.

2b. Glumes acuminate-cuspidate, 2–2.5 mm long, more than half as long as the spikelet; culms up to 30 cm long, densely tufted from hard bulb-like scaly bases; inflorescence 5–10 cm long. *Muhlenbergia cuspidata* (Torr.) Rydb.; prairie muhly. Shale bank prairie on slope near East Gate; rare.

3a. Lemma, with an awn up to 10 mm long; hairs at base of lemma copious, as long as the lemma; glumes awnless or awn-tipped; culms up to 60 cm long, from elongated, wiry, and scaly rhizomes; inflorescence 7–15 cm long, spike-like. *Muhlenbergia andina* (Nutt.) Hitchc.; foxtail muhly. Gravelly shore, Clear Lake; rare.

3b. Lemmas awnless; hairs at base of lemma not conspicuous, usually less than half as long as the lemma; glumes awnless, awn-tipped or awned . . . . (4)

4a. Glumes awnless or awn-tipped, about as long as the lemma; culms up to 60 cm long, from creeping scaly rhizomes; panicle 10–15 cm long, dense. *Muhlenbergia mexicana* (L.) Trin.; wood muhly. Gravel pit in rolling scrub prairie; rare.

4b. Glumes awned, much longer than the lemmas . . . . (5)

5a. Leaf sheath keeled; ligule 1–1.5 mm long; culms up to 50 cm long, from creeping scaly rhizomes, and usually branching from the middle nodes; internodes smooth; anthers 0.5–0.8 mm long. *Muhlenbergia racemosa* (Michx.) BSP; marsh muhly. Disturbed situations; occasional.

5b. Leaf sheath not keeled, ligule minute; culms up to 50 cm long, from long, branching, scaly rhizomes, and usually simple or branching from the base; internodes puberulent; anthers 1.0–1.5 mm long. *Muhlenbergia glomerata* (Willd.) Trin.; bog muhly. Bogs, scrub prairie, open woodland; occasional.

Fig. 64. *Muhlenbergia richardsonis*, 3×.

Oryzopsis   rice grass, mountain rice

1a. Leaves flat, up to 10 mm wide, often overtopping the inflorescence, evergreen; culms often purplish at the base, up to 70 cm long; panicle narrow, 5–10 cm long, with ascending spikelets; lemmas 7–9 mm long, densely pubescent at the base, with 5–10 mm long awns. ***Oryzopsis asperifolia*** Michx.; winter grass, mountain rice; Fig. 65. Drier woodlands in humus or lighter soils; common.

1b. Leaves involute or filiform .... (2)

2a. Panicle 5–10 cm long, open, lax; branches flexuous, ascending, or spreading; culms up to 60 cm long; leaf blades 2–4 mm wide, flat to involute; lemmas about 3 mm long, appressed pubescent; awn 1–2 cm long, weakly twice geniculate. ***Oryzopsis canadensis*** (Poir.) Torr.; Canada rice grass. Scrub prairie and open mixed woods; rare.

2b.   Panicle 3–6 cm long; branches ascending-appressed or spreading; culms up to 60 cm long; leaves filiform, stiff, and scabrous; lemmas 3–4 mm long, densely pubescent; awns 1–3 mm long. ***Oryzopsis pungens*** (Torr.) Hitchc.; Fig. 66. In open sometimes disturbed lighter soils on the east side of the park; rare.

Phalaris   canary grass

Leafy culms up to 150 cm long or longer; rhizomes stout, horizontally creeping; leaf blades flat, up to 25 cm long and 1.5 cm wide; panicle up to 20 cm long, open in anthesis but tightly contracted in fruit; spikelets lanceolate, laterally flattened. ***Phalaris arundinacea*** L.; reed canary grass; Fig. 67. Disturbed road allowances, waste places, lakeshores, and stream banks, both native and introduced; frequent.

Phleum   timothy

Culms up to 80 cm long, often forming large clumps from a swollen bulb-like base; panicles dense, firm, cylindrical, spike-like. ***Phleum pratense*** L.; timothy. Introduced forage grass, frequent along roadsides and in waste places.

Phragmites   reed

Our tallest grass; leafy culms bamboo-like, up to 2 m long; rhizomes long, thick, extensively creeping, sometimes forming extensive colonies; leaves broad and flat; panicles plumose, up to 40 cm long; lemmas deep purple, long-acuminate. ***Phragmites australis*** (Cav.) Steud. (*P. communis* Trin.); reed; Fig. 68. Wet lakeshores and ditches; localized and very distinctive.

Poa   blue grass, meadow grass

1a.   Annual; culms up to 20 cm long, tufted, soft, divergent, decumbent or prostrate; panicle open, pyramidal, usually pale green; branches mainly in twos. ***Poa annua*** L.; annual blue grass. Introduced weedy species found on moist shores, in disturbed situations, and in lawns; frequent.

1b.   Perennial; tufted or rhizomatous . . . . (2)

2a.   Plants rhizomatous . . . . (3)

Fig. 65. *Oryzopsis asperifolia*, 1/4×.

Fig. 66. *Oryzopsis pungens*, 2×.

Fig. 67. *Phalaris arundinacea*, 2/5×.

2b.   Plants in dense clumps or tussocks, not rhizomatous
. . . . (5)

3a.   Culms up to 60 cm long, flattened, especially the lower part; sheaths strongly keeled above the middle; panicle short, 3–10 cm long; branches mostly in twos. ***Poa compressa*** L.; Canada blue grass, wire grass. Introduced weedy species usually found in dry disturbed soils; occasional.

3b.   Culms terete; sheaths not strongly keeled . . . . (4)

4a.   Panicle lanceolate, thick; lemmas not cobwebby at the base, pubescent to pilose dorsally towards the base on the internodes; culms up to 50 cm long, solitary or in small clusters. ***Poa arida*** Vasey; plains blue grass. Scrub prairie and disturbed situations; occasional.

4b.   Panicle widely open; lemmas cobwebby at the base, glabrous or slightly pubescent toward the base on the nerves only; culms up to 100 cm long, tufted, with rhizomes. ***Poa pratensis*** L.; Kentucky blue grass; Fig. 69. Common lawn grass, prairies, clearings, and roadsides.

5a.   Culms in dense tufts, up to 70 cm long; ligules of stem leaves truncate, mostly less than 1.5 mm long; panicles up to 10 cm long, loose, often nodding; branches spreading in flower, later appressed. ***Poa nemoralis*** L. (incl. *P. interior* Rydb.); wood blue grass. Meadows and open woods; occasional.

5b.   Culms in loose tufts, up to 100 cm long; ligules of stem leaves 2–4 mm long or longer, somewhat acute at the tip; panicles up to 20 cm long or longer, pyramidal or oblong; branches spreading. ***Poa palustris*** L.; fowl blue or meadow grass; Fig. 70. Wet meadows, grassy shores, and ditches; frequent.

Puccinellia   alkali grass

Tufted perennial; culms erect to decumbent or geniculate-ascending, up to 70 cm long; panicles up to 15 cm long, pyramidal; branches reflexed at maturity. ***Puccinellia distans*** (L.) Parl.; alkali grass. Roadsides; rare, introduced from Europe.

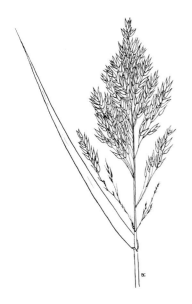

Fig. 68. *Phragmites australis*, 2/5×.

Fig. 70. *Poa palustris*, 1/4×.

Fig. 69. *Poa pratensis*, 2/5×.

Schizachne   purple oat grass

Loosely tufted rhizomatous species; culms up to 100 cm long; panicles up to 15 cm long, often secund; branches more or less drooping; spikelets divergently aristate; glumes reddish purple; margins green, hyaline; florets light green to purplish at the tip. *Schizachne purpurascens* (Torr.) Swallen; purple oat grass; Fig. 71. Open woodland and scrub prairie; frequent.

Scolochloa   spangletop

Culms stout, leafy, erect, up to 1.5 m long, spongy at the base, from extensively creeping whip-like rhizomes; leaf blades flat, up to 10 mm wide; sheaths large, papery, inflated; panicles open, up to 20 cm long; branches in distant fascicles; spikelets about 8 mm long, 3 or 4 flowered. *Scolochloa festucacea* (Willd.) Link; spangletop; Fig. 72. In shallow water of marshes and bordering ponds where it may form dense stands.

Setaria   foxtail

Annual, tufted; culms erect or geniculate-ascending, up to 50 cm long or longer; leaf blades up to 12 mm wide; panicles up to 10 cm long and 2 cm wide (including the awns). *Setaria viridis* (L.) Beauv.; green foxtail, bottle grass. Weed of waste places.

Spartina   cord grass

Stems up to 100 cm long, solitary or in small tufts from tough rhizomes; spikes 4–8, 1-sided; spikelets crowded in 2 rows along the edges. *Spartina gracilis* Trin.; alkali cord grass. Disturbed area near Rennicker Creek; apparently localized.

Sphenopholis   wedge grass

Tufted perennial; culms slender, up to 70 cm long; leaf blades flat, up to 5 mm wide; panicles up to 20 cm long, dense, and appearing lobed; spikelets densely crowded on short ascending branches; glumes subequal, with the first one linear-subulate and the second obovate. *Sphenopholis intermedia* Rydb.; slender wedge grass; Fig. 73. Wet meadows, lakeshores, and stream banks; occasional.

Fig. 71. *Schizachne purpurascens*, 4/5 ×.

Fig. 73. *Sphenopholis intermedia*, 1/2 ×.

Fig. 72. *Scolochloa festucacea*, 1/4 ×.

Sporobolus   dropseed

Culms slender, erect, up to 70 cm long, forming thick compact tufts; leaves flat to involute, erect or slightly drooping, often almost as long as the culms; panicles open, up to 20 cm long, narrowly ovoid, with the lower part often enclosed in the upper sheath; spikelets deep green to blackish, longer than their pedicel and crowded on the branches. ***Sporobolus heterolepis*** A. Gray; prairie dropseed. Scrub prairie; rare.

Stipa   spear grass; needle grass

1a.   Panicles up to 20 cm long; branches widely spreading, with only a few drooping spikelets towards the ends; tufted plants with narrow, involute leaves about 20 cm long and culms up to 80 cm long; awn 2.5–3.5 cm long, once geniculate. ***Stipa richardsonii*** Link; needle grass; Fig. 74. Scrub prairie; frequent or locally abundant.

1b.   Panicles closed and narrow; branches shorter . . . . (2)

2a.   Glumes 20–30 mm long; lemmas up to 15–25 mm long; awns up to 12 cm long, twice geniculate; culms up to 80 cm long or longer; forming dense tussocks; panicles 15–20 cm long; branches each bearing 1 or 2 spikelets. ***Stipa spartea*** Trin. var. ***curtiseta*** Hitchc.; western porcupine grass, spear grass; Fig. 75. Scrub prairie; frequent or locally abundant.

2b.   Glumes 8–10 mm long; lemmas 5–6 mm long; awns 2–3 cm long, twice geniculate and spirally twisted below; culms up to 100 cm long, loosely tufted; panicles up to 20 cm long; branches appressed-ascending, bearing 1–7 spikelets. ***Stipa viridula*** Trin.; feather bunch grass, green needle grass. Scrub prairie; rare.

Torreyochloa

Culms up to 60 cm long, geniculate, loosely matted; panicles up to 13 cm long; lower branches strongly divergent or reflexed in age; in aspect like a short-spikelet *Glyceria*, but the sheath margins are free and the upper glume has 3 nerves. ***Torreyochloa pallida*** (Torr.) Church var. ***fernaldii*** (Hitchc.) Dore (*Glyceria fernaldii* (Hitchc.) St. John). Wet bottom of catchbasin on east slope; rare.

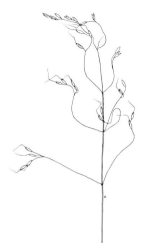

Fig. 74. *Stipa richardsonii*, 1/4×.

Fig. 75. *Stipa spartea* var. *curtiseta*, 2/5×.

Triticum   **wheat**

1a.   Keel of the glume winged near the tip only; tufted annual; culms up to 1 m long; leaves flat up to 2 cm wide; spikes fat, up to 12 cm long; usually awnless. ***Triticum aestivum*** L.; soft wheat. A common cultivated crop appearing as a volunteer along roadsides.

1b.   Keel of the glume produced into a narrow wing for its whole length; in aspect much like *T. aestivum*, but usually with long, coarse, rough awns. ***Triticum turgidum*** L.; hard wheat. Occasional spontaneous plant along roadsides.

16. CYPERACEAE   sedge family

1a.   Flowers unisexual, with the staminate and pistillate in the same or in different spikes; achene enclosed in a sac (perigynium); culm sharply or obtusely 3-angled .... ***Carex***

1b.   Flowers all perfect, in uniform spikelets; achenes naked .... (2)

2a.   Base of style persistent as a tubercule at the summit of the achene; culms leafless .... ***Eleocharis***

2b.   Base of style not persistent as a tubercule .... (3)

3a.   Perianth bristles numerous, often 2–3 cm long, silky .... ***Eriophorum***

3b.   Perianth bristles 1–8, occasionally lacking, shorter than or only slightly longer than the achene .... ***Scirpus***

Carex   **sedge**

1a.   Spike solitary, bractless .... (2)

1b.   Spikes 2 or more .... (4)

2a.   Achenes lenticular; stigmas 2; perigynia glabrous; spike wholly staminate, or wholly pistillate, or staminate above and pistillate below; culms up to 30 cm, solitary or few, from slender creeping rhizomes. ***Carex gynocrates*** Wormsk.; Fig. 76. *Sphagnum* bogs; localized.

2b.   Achenes triangular in cross section; stigmas 3 .... (3)

3a.   Staminate scales with margins united nearly to the middle; perigynia subalternate, beakless; leaves very lax and soft; culms up to 40 cm long, flaccid, densely caespitose to substoloniferous. ***Carex leptalea*** Wahl.; Fig. 77. Moist moss, usually in spruce woods; localized.

3b.   Staminate scales with margins free to the base; leaves firm, flat; perigynia castaneous to blackish, lustrous; culms solitary or few together, from tough cord-like rhizomes. ***Carex obtusata*** Lilj.; Fig. 78. Dry grassland hilltops; locally common.

4a.   Achenes lenticular or plump, not triangular in cross section; stigmas 2 .... (5)

4b.   Achenes trigonous (triangular in cross section); terminal spike mostly entirely staminate, sometimes partly pistillate; at least some of the lateral spikes strictly pistillate, mostly peduncled .... (31)

5a.   Spikes of 2 kinds, with the terminal ones staminate (at least below) and the lower ones entirely or mostly pistillate, peduncled .... (6)

5b.   Spikes essentially uniform, with the lateral ones sessile .... (7)

6a.   Bracts sheathing; perigynia golden yellow at maturity, beakless; culms up to 30 cm long, usually spreading or ascending, from slender creeping rhizomes. ***Carex aurea*** Nutt.; Fig. 79. Meadows, damp woods, ditches, and shores; occasional.

6b.   Bracts nearly or quite sheathless; perigynia 2–4 mm long, purplish green or green, minutely beaked; culms up to 80 cm long, densely tufted, often in large clumps, with long scaly rhizomes; pistillate spikes 2–4 cm long, usually 2–6. ***Carex aquatilis*** Wahl.; Fig. 80. Fens, sloughs, and wet meadows; frequent.

7a.   Culms arising singly or few together from long-creeping slender or cord-like rhizomes; spikes androgynous (except *C. siccata*, with terminal spike usually pistillate above a prolonged staminate base) .... (8)

7b.   Culms densely to loosely caespitose; rhizomes sometimes short-prolonged, but not long-creeping .... (10)

8a. Culms branching, with the long prostrate ones of the previous year bearing erect culms from the axils of the old, dried-up leaves; leaves narrow, involute; spikes few, crowded. **Carex chordorrhiza** L.f.; Fig. 81. Bogs; localized.

8b. Culms simple .... (9)

9a. Perigynia 2.5–4.5 mm long; beaks one-quarter to one-third as long as the body; inner band of leaf sheaths green-nerved nearly to the summit; culms up to 80 cm long, sharply triangular, stiff; upper nodes exserted from the sheaths; rhizomes slender, black. **Carex sartwellii** Dewey; Fig. 82. Low, moist areas; apparently infrequent.

9b. Perigynia 5–6 mm long; margins narrow, greenish; beaks two-thirds as long as the body; leaf sheaths much overlapping, with their inner bands nerveless; culms up to 80 cm long, solitary or few together from long, tough, brown rhizomes. **Carex siccata** Dewey (*C. foenea* Willd.). Usually dry lighter soils in the open; localized.

10a. Spikes androgynous (*C. disperma* may be found here) .... (11)

10b. Spikes, at least the terminal, gynaecandrous .... (17)

11a. Culms soft and flattening in drying, serrulate above, with concave sides and with sharp angles more or less winged; leaf sheaths loose .... (12)

11b. Culms relatively firm, not flattening in drying; leaf sheaths close .... (13)

12a. Perigynia ovate, 3–4 mm long, stipitate, flat on the inner nerveless face, faintly brown-nerved dorsally, contracted into a beak about half as long as the body, about equalling or little exceeding the brownish scales; culms up to 70 cm long, tufted, mostly as long as or shorter than the leaves; inner band of leaf sheaths dotted, unwrinkled; upper spikes crowded, with the lower ones distant. **Carex alopecoidea** Tuck. Borders of lakes and streams and in wet meadows; occasional.

12b. Perigynia subulate-lanceolate, 4–5 mm long, brown-nerved, much exceeding the pale scales, tapering from a broad base to a long slender beak; culms up to 100 cm long, growing from short thick rhizomes; inner band of leaf sheaths commonly cross-puckered, at least between the nerves; inflorescence of 5–15 spikes aggregated into a head,

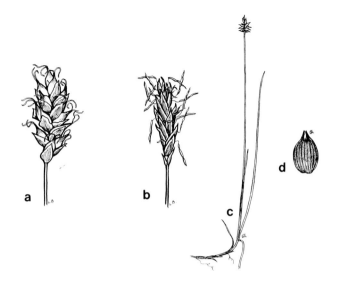

Fig. 76. *Carex gynocrates, a,* 2×; *b,* 2×; *c,* 1/2×; *d,* 4×.

Fig. 77. *Carex leptalea,* 4×.

Fig. 78. *Carex obtusata,* 2 1/3×.

77

with the upper ones crowded together. **Carex stipata** Muhl. Swamps, borders of creeks and ponds, and wet open woodland; occasional.

13a. Spikes 3–10, in simple, close, or interrupted heads; perigynia nerveless or lightly striate dorsally; sheaths not red-dotted ventrally .... (14)

13b. Spikes numerous in a more or less compound head, usually 2 to several spikes on each lateral branch; perigynia dorsally nerved .... (15)

14a. Perigynia of nearly uniform texture; margins not incurved; bracts with brown, broad scarious bases; scales pale brown, strongly awned, covering the perigynia; culms up to 40 cm long, from short rhizomes; spikes 5–8, distant to approximate. **Carex hookeriana** Dewey. Scrub prairie; localized.

14b. Perigynia corky-thickened below the middle; nerve-like margin inflexed; scales whitish, rounded, or somewhat obtuse; culms weak and lax, up to 40 cm long or longer, densely tufted; spikes remote; perigynia spreading horizontally. **Carex rosea** Schkuhr. Moist woods and clearings; rare.

15a. Scales long-awned; bracts setaceous, conspicuous; perigynia flat on the inner face, yellowish green or straw-colored; culms up to 80 cm long, densely tufted; inner band of leaf sheaths cross-puckered, not red-dotted; inflorescence of 4-8 compound densely crowded spikes. **Carex vulpinoidea** Michx. Stream banks and lakeshores; rare.

15b. Scales not awned; bracts mostly short and inconspicuous or wanting; perigynia straw-colored to dark brown; inner nerveless band of leaf sheaths often red-dotted, not cross-puckered .... (16)

16a. Sheaths strongly copper-tinged at the mouth; perigynia flat on the inner face, dull brownish in age, appressed, covered by the scales; culms up to 100 cm long, densely tufted, from short stout rhizomes; inflorescence lax and open. **Carex prairea** Dewey. Swamps; localized.

16b. Sheaths not copper-tinged at the mouth; perigynia slightly convex ventrally, lustrous, blackish in age, soon divergent, mostly longer than the scales; culms up to 80 cm long, forming dense tufts, from short rhizomes; inflorescence of 6–10 spikes, with the lower

Fig. 81. *Carex chordorrhiza*, *a*, 1/2×; *b*, 2×.

Fig. 79. *Carex aurea*, 4/5×.

Fig. 80. *Carex aquatilis*, 3/5×.

ones somewhat distant and the upper ones approximate to crowded. **Carex diandra** Schrank; Fig. 83. Swamps and fens; localized.

17a. Perigynia at most thin-edged, corky-thickened at the base; culms not hollow . . . . (18)

17b. Perigynia narrowly to broadly wing-margined, not spongy-thickened at the base, beaked, mostly with concave inner faces; culms hollow . . . . (25)

18a. Perigynia closely appressed-ascending, 4.5–5.5 mm long; beaks long, slender, bidentate; scales pale, acuminate or short-cuspidate; plants densely tufted; culms up to 80 cm long, weak; leaves glaucous; inflorescence 3–8 cm long; spikes 3–5, with the lowest one having a filiform bract up to 3 cm long. **Carex deweyana** Schw.; Fig. 84. Moist woodland, clearings, and ditches; occasional.

18b. Perigynia 1.5–3.75 mm long . . . . (19)

19a. Perigynia white-puncticulate, ascending or merely spreading-ascending, soft in texture (except in *C. disperma*), nearly beakless or with a short subentire beak and with margins rounded . . . . (20)

19b. Perigynia not puncticulate, spreading or reflexed at maturity, beaked, very spongy at the base and with margins thin . . . . (24)

20a. Staminate flowers terminal in the spikes; perigynia nearly round in cross section, 2.0–2.8 mm long, finely many-nerved, rounded at the summit to a minute entire beak; culms up to 50 cm long, very slender, weak, and spreading from slender, creeping sod-forming rhizomes; spikes 2–5, remote. **Carex disperma** Dewey; Fig. 85. Bogs and wet woods; frequent.

20b. Staminate flowers basal or scattered; perigynia flattened on 1 face . . . . (21)

21a. Bract at base of inflorescence bristle-like and much prolonged, much longer than the subtended spike; inflorescence flexuous; spikes 2 or 3, widely separated; plants loosely caespitose; culms weak, usually overtopping the 1–2 mm-wide leaves. **Carex trisperma** Dewey. Marginal swamp; apparently rare and localized but perhaps overlooked.

21b. Bract at base of inflorescence shorter or none; inflorescence straightish; at least the upper spikes approximate . . . . (22)

Fig. 82. *Carex sartwellii*, 1 ×.

Fig. 84. *Carex deweyana*, 1 2/5 ×.

Fig. 83. *Carex diandra*, 1 2/5 ×.

Fig. 85. *Carex disperma*, 1 2/5 ×.

22a.   Perigynia beakless or nearly so; scales white; spikes 2–4, subglobose, closely approximate in an ellipsoid or subglobose head; plants loosely caespitose; culms subcapillary, firm, mostly longer than the leaves. **Carex tenuiflora** Wahl. Marginal swamp; apparently rare and localized but perhaps overlooked.

22b.   Perigynia definitely beaked . . . . (23)

23a.   Perigynia 5–10 in each spike, loosely spreading at maturity; serrulate at the base of the distinct beak; plants green; culms up to 70 cm long, densely tufted; leaves much shorter than the culms; inflorescence with 5–8 short ovoid or subglobose spikes, with the lower ones distant and the upper ones approximate. **Carex brunnescens** (Pers.) Poir.; Fig. 86. Bogs and wet woods; rare.

23b.   Perigynia 10–30 in each spike, appressed, smooth, or at most sparsely serrulate at the base of the inconspicuous beak; plants glaucous; culms up to 25 cm long, often recurved or curved ascending, loosely tufted from slender rhizomes; inflorescence of 4–6 spikes, with the lower ones distant and the upper ones approximate. **Carex curta** Good. (*C. canescens* L.); Fig. 87. Bogs and wet woods; rare.

24a.   Perigynia nerveless or weakly few-nerved at the base ventrally; beak shallowly bidentate, one-quarter to one-third the length of the body; plants densely tufted; culms slender, up to 70 cm long; inflorescence usually consisting of 3 small often remote spikes, of which the terminal is clavate; perigynia spreading at maturity. **Carex interior** Bailey. Bogs, lakeshores, and wet woods; frequent.

24b.   Perigynia slightly nerved on both faces; beak sharply bidentate, about half as long as the body; plants densely caespitose, from short rhizomes; culms up to 80 cm long; spike solitary (occasionally with 1–3 accessory spikes below the main one), pistillate above and staminate below or dioecious; terminal spike without a distinct clavate base of staminate scales. **Carex sterilis** Willd. Open calcareous fen; rare.

25a.   Bracts leaf-like and many times exceeding the dense inflorescence; perigynia lance-subulate, about 5 mm long; scales smaller than the perigynia, hyaline with a green midrib; plants densely tufted; culms up to 50 cm long. **Carex sychnocephala** Carey; Fig. 88. Creek margins, catchbasins, and ditches; occasional.

Fig. 86. *Carex brunnescens*, 1 2/5×.

Fig. 87. *Carex curta*, 1 2/5×.

Fig. 88. *Carex sychnocephala*, 1 2/5×.

25b.   Bracts (when present) neither leaf-like nor prolonged
. . . . (26)

26a.   Scales about equaling the perigynia and nearly or
completely covering them . . . . (27)

26b.   Scales shorter than the perigynia and noticeably
narrower above . . . . (29)

27a.   Inflorescence dense and erect, consisting of crowded or
approximate spikes; inner face of perigynia nerveless
or nerved only at the base . . . . (28)

27b.   Inflorescence interrupted and flexuous, with at least
the lower spikes remote; scales brown; perigynia
4.5–6.5 mm long, not closely appressed; inner face
nerveless or only short-nerved; plants densely tufted,
from short rhizomes; culms up to 80 cm long. **Carex
praticola** Rydb.; Fig. 89. Meadows, open woods, and
clearings; occasional.

28a.   Perigynia thin and scale-like, 4.0–4.8 mm long, closely
appressed; inner face nerveless or nerved only at the
base; scales light brown; midrib green; margins
hyaline; plants densely tufted, from short thick
rhizomes; culms up to 60 cm long. **Carex xerantica**
Bailey. Fescue prairie; rare.

28b.   Perigynia plump, 4–5 mm long, loosely ascending in
age; scales brown; margins hyaline; plants tufted;
culms up to 60 cm long, smooth. **Carex adusta** Boott.
Moist situations; rare.

29a.   Beak of perigynium slender and roundish in cross
section, little if at all serrulate at the scarcely
margined tip; perigynia thin, 3.75–5 mm long; scales
dark brown; inflorescence dense; plants densely
tufted, from short rhizomes; culms up to 40 cm long.
Carex microptera Mack. (*C. festivella* of Scoggan).
Stream banks and borders of lakes; rare.

29b.   Beak of perigynium flattened and margined to the
serrulate tip; perigynium firm, 1–2 mm wide, broadly
cuneate or rounded at the base . . . . (30)

30a.   Inflorescence compact, 1–4 cm long; plants densely
tufted; culms up to 60 cm long or longer, commonly
equalling the leaves in length. **Carex bebbii** Olney ex
Fern.; Fig. 90. Ditches, streambanks, and borders of
ponds; frequent.

30b.   Inflorescence interrupted, 2–5 cm long, flexuous or
nodding; plants densely tufted and producing
numerous tall sterile shoots; culms up to 60 cm long.

Fig. 89. *Carex praticola*, 1×.   Fig. 90. *Carex bebbii*, 1 2/5×.

Carex tenera Dew. Damp open woodland and borders of clearings; occasional.

31a.   Staminate scales very tight, with margins united at the base; lower pistillate scales green and bract-like; perigynia 4.0–5.6 mm long, tapering to a conical beak; plants densely caespitose; culms up to 35 cm long; leaves deep green, flat, up to 6 mm wide, often longer than the culms. **Carex backii** Boott. Open woodland and borders of clearings; rare.

31b.   Staminate scales with margins free to the base, loosely ascending; pistillate scales not bract-like . . . . (32)

32a.   Perigynia pubescent . . . . (33)

32b.   Perigynia glabrous or essentially so . . . . (43)

33a.   Bract at base of inflorescence long-sheathing . . . . (34)

33b.   Bract at base of inflorescence sheathless or nearly so
. . . . (36)

34a.   Basal sheath of inflorescence with a long leaf-like blade; scales greenish straw-colored, lance-acuminate or short-awned; perigynia lance-subulate, 5.0–6.5 mm long; plants loosely tufted; culms up to 60 cm long; inflorescence up to 20 cm long; terminal spike staminate; pistillate spikes narrow, remote, on long leafy bracted peduncles. **Carex assiniboinensis** Boott. Border of mixed woods; rare.

34b. Basal sheath of inflorescence bladeless; scales dark, pale-margined .... (35)

35a. Scales exceeding the perigynia; pistillate spikes cylindrical, 1–2 cm long; staminate spike 1.5–2.5 cm long, often stalked; plants tufted, growing from long rhizomes; culms up to 25 cm long; bladeless sheaths strongly purple-tinged. **Carex richardsonii** R. Br. Dry open woods and clearings; rare.

35b. Scales roundish, much shorter than the perigynia; pistillate spikes subglobose, 4–7 mm long; staminate spikes 3–6 mm long, often sessile; plants loosely tufted, growing from slender rhizomes; culms up to 35 cm long, slender, erect to curved; leaves often as long as the culms. **Carex concinna** R. Br.; Fig. 91. Moist moss in open spruce woods; occasional.

36a. Achenes only obscurely 3-angled; sides rounded or convex .... (37)

36b. Achenes definitely 3-angled; sides flat or concave .... (41)

37a. Culms all elongate .... (38)

37b. Culms of various lengths, with the shorter ones crowded among the leaf bases; spikes often all pistillate .... (40)

38a. Body of perigynium subglobose, somewhat loose over the achene; scales nearly equalling or exceeding the perigynia; terminal staminate spike 0.8–2.0 mm long; plants tufted, from extensively creeping, slender rhizomes; culms up to 30 cm long, usually with persistent brush-like tufts of fibers at the base and usually exceeding the leaves. **Carex pensylvanica** Lam. Prairie openings and dry open woodland; occasional.

38b. Body of perigynium ellipsoid to fusiform-obovoid, definitely longer than thick, tightly investing the achene .... (39)

39a. Perigynia 3–4 mm long, copiously short-hirsute, gradually narrowed to a thick spongy base; scales with broad white margins; plants forming a loose carpet from extensive cord-like rhizomes; culms up to 40 cm long with reddish bases, overtopping the leaves; inflorescence usually crowded. **Carex peckii** Howe. Mixed woods; frequent.

Fig. 91. *Carex concinna*, 2×.

39b. Perigynia 2.5–3.0 mm long, minutely pubescent, gradually narrowed to a long slender stipe; scales with purple margins; plants tufted, from stout horizontal or ascending rhizomes; culms up to 20 cm long, flexuous, mostly shorter than the deep green leaves; bases purplish; inflorescence crowded. **Carex deflexa** Hornem.; Fig. 92. Moist open woods; rare.

40a. Perigynia 3.0–4.5 mm long; beaks 0.7–1.5 mm long; staminate spike 3–15 mm long; bract at the base of the inflorescence 0.5–5 cm long; plants often in thick clumps, but with slender rhizomes; culms erect, harsh above, up to 30 cm long, often exceeded by the erect or curled leaves. **Carex rossii** Boott.; Fig. 93. Usually dry open disturbed situations; occasional.

40b. Perigynia 2.5–3.0 mm long; beaks about 0.5 mm long; staminate spike 2–5 mm long; basal bract 5–10 mm long; leaves soft; culms flexuous, smooth except at the tip. See *Carex deflexa*.

41a. Perigynia strongly 15–20-ribbed; beaks nearly half as long as the body; teeth spreading; plants loosely tufted from long-creeping thick rhizomes; culms up to 60 cm long; leaves up to 1 cm wide; inflorescence up to 15 cm long; terminal spike staminate; pistillate spikes 1–3, remote, either sessile or short-petioled, leafy-bracted. **Carex houghtoniana** Torr. (*C. houghtonii* Torr.). Dry to moist sandy shores and clearings; rare.

41b. Perigynia with obscure ribs mostly concealed beneath the dense pubescence; beaks short; teeth erect ... (42)

42a. Leaves flat, 2–5 mm wide, scabrous; margins revolute; culms sharply triangular and scabrous above, up to 70 cm long, usually in small tufts, from slender long-creeping rhizomes; inflorescence up to 20 cm long; terminal spikes 1 or 2, staminate; pistillate spikes 1–3, remote, sessile or short-peduncled; perigynia 2.5–3.5 mm long, densely pubescent, abruptly beaked; erect teeth about 0.5 mm long. **Carex lanuginosa** Michx. Swamps, ditches, and borders of ponds and streams; frequent.

42b. Leaves filiform-convolute except at the base, 2 mm wide or less, smooth and wiry; culms up to 120 cm long, obtusely angled and smooth except sometimes at the tip, from stout long-creeping rhizomes; inflorescence up to 35 cm long; terminal spikes 1–3, staminate; pistillate spikes 2–3, remote; perigynia 4–5 mm long, densely pubescent, tapering into the beak; teeth about 1 mm long. **Carex lasiocarpa** Ehrh. var. **americana** Fern. Fens; localized.

43a. Style continuous with the achene and of the same bony texture, not withering; leaves more or less septate-nodulose .... (44)

43b. Style jointed to the achene, not indurated, soon disarticulating and shriveling .... (49)

44a. Perigynia firm, scarcely inflated; culms 1 to few, from stoloniferous bases .... (45)

44b. Perigynia thin or papery, often strongly inflated .... (46)

45a. Teeth of perigynia 1 mm long or less; perigynia lightly many-nerved; plants tufted, from long rhizomes; culms up to 125 cm long, sharply triangular with rough edges; inflorescence up to 35 cm long; terminal spikes 2–4, staminate, 1–8 cm long; pistillate spikes 2–4, usually distinct, 3–10 cm long, erect, sessile or short-peduncled. **Carex lacustris** Willd. Marshes and fens; localized.

Fig. 92. *Carex deflexa*, 2×.   Fig. 93. *Carex rossii*, 1 2/5×.

45b.   Teeth of perigynia 1.6–3 mm long, outwardly curving;
        perigynia strongly ribbed; plants loosely tufted, from
        creeping rhizomes; culms up to 120 cm long; sheaths,
        especially the lower ones, pubescent; terminal spikes
        2–6, staminate, 2–4 cm long; pistillate spikes 2–4, 4–10
        cm long, remote, erect, sessile or short-peduncled.
        Carex atherodes Spreng.; Fig. 94. Fens, lakeshores,
        and stream banks; common.

46a.   Pistillate scales with midrib excurrent as a long,
        scabrous awn; perigynia finely and closely ribbed,
        several times longer than the body of the scales
        . . . . (47)
46b.   Pistillate scales blunt to short-awned; perigynia
        mostly inflated, coarsely ribbed . . . . (48)

47a.   Perigynia 3–5 mm long, soon strongly reflexed, more
        or less 2-edged, not inflated; beak shorter than the
        body; sod-forming plant with short rhizomes; culms to
        1 m, sharply 3-angled, scabrous, nodding at the top;
        inflorescence up to 20 cm long; terminal spike
        staminate, 2–5 cm long; pistillate spikes 3–5, 3–7 cm
        long, with the upper ones approximate and the lower
        ones remote, spreading or drooping on slender
        peduncles; **Carex pseudo-cyperus** L. Fens and
        lakeshores; rare.

47b. Perigynia about 6 mm long, spreading, inflated; beak slender, as long as the body; densely tufted plant; erect culms up to 70 cm long; inflorescence up to 15 cm long; terminal spike staminate, 2–4 cm long; pistillate spikes 1–4, up to 4 cm long, with the upper ones approximate and the lowest ones often remote, peduncled, erect to spreading. *Carex hystricina* Muhl. Lakeshores and wet clearings; rare.

48a. Perigynia 7–10 mm long, soon reflexed or horizontally spreading, much exceeding the acuminate scales; culms densely tufted, obtusely 3-angled up to 100 cm long, from short stout rhizomes; inflorescence up to 15 cm long; upper 1–4 spikes staminate, often hidden among the 3–8, 1.5–6.0-cm-long, aggregated pistillate spikes; lower bracts much prolonged, many times the length of the inflorescence. *Carex retrorsa* Schw.; Fig. 95. Ditches, fens, borders of lakes and streams; frequent.

48b. Perigynia ascending to merely spreading; plants with short rhizomes and long stolons; culms stout, up to 100 cm long, usually exceeded by the leaves; inflorescence up to 30 cm long; upper 2–4 spikes staminate, to 5 cm long; pistillate spikes 2–5, well separated, 4–10 cm long, sessile to short-peduncled; bracts not more than a few times longer than the inflorescence. *Carex rostrata* Stokes; Fig. 96. Fens, swamps, and wet stream banks and lake margins; common.

49a. Bract at the base of the inflorescence sheathless or very short-sheathing . . . . (50)
49b. Bract at the base of the inflorescence with a prolonged tubular sheath . . . . (52)

50a. Leaves short-pilose; culms up to 50 cm long, slender, weak, sharply triangular, from short rhizomes; inflorescence up to 5 cm long; terminal spike staminate; pistillate spikes 1–3, approximate, up to 15 mm long, sessile or short peduncled; perigynia 2.5–3.5 mm long, yellowish green, strongly nerved, abruptly short-beaked. *Carex torreyi* Tuck. Open woodland and scrub prairie; frequent.
50b. Leaves glabrous . . . . (51)

51a. Plants strongly rhizomatous; rhizomes covered with a brownish felt-like layer; culms up to 50 cm long, sharply 3-angled; inflorescence up to 6 cm long; terminal spike staminate, 1–2 cm long; pistillate 1–2

Fig. 95. *Carex retrorsa*, 2/5×.

Fig. 94. *Carex atherodes*, 2/5×.

Fig. 96. *Carex rostrata*, 2/5×.

spikes to 15 mm long, drooping on filiform peduncles; perigynia 2.5–4.0 mm long, 8–10-nerved; scales brown, obtuse to acute, with a green midrib about as long as the perigynia. **Carex limosa** L.; Fig. 97. Quaking fen; rare.

51b.   Plants loosely caespitose; rootlets covered with a yellow felt; culms up to 60 cm long, slender; inflorescence up to 12 cm long; terminal spike staminate, 5–15 mm long; pistillate spikes 2–4, 8–20 mm long, on slender peduncles, with the upper approximate and the lower remote; perigynia 3.0–3.5 mm long, veinless to finely veined; scales long-acuminate, as long as or longer than the perigynia, but narrower. **Carex magellanica** Lam. (*C. paupercula* Michx.); Fig. 98. Sphagnum moss under black spruce; rare.

52a.   Sheaths spathe-like, bladeless or very short-bladed; plants forming dense mats from short thick rhizomes; culms up to 30 cm long, brownish at the base, barely exceeding the leaves; inflorescence 4–10 cm long; terminal spike staminate 5–15 mm long; pistillate spikes 2–3, up to 30 mm long, with the upper ones ascending and the lower ones remote, on long pedicels; perigynia 3.5–5 mm long, pubescent; scales smaller, purplish brown, with a green midrib extended into a short awn. **Carex pedunculata** Muhl. Damp white spruce woodland; rare.

52b.   Sheaths with well developed blades . . . . (53)

53a.   Leaves and culms pilose; culms up to 70 cm long, tufted, purplish at the base; inflorescence up to 10 cm long; terminal spike staminate, 1–2 cm long; pistillate spikes 2–3, up to 2 cm long or longer, spreading or drooping on slender peduncles; perigynia 4–6 mm long, with a slender bidentate beak; scales brownish about equaling the perigynia. **Carex castanea** Wahl. Damp open spruce woods and clearings; occasional.

53b.   Leaves and culms glabrous . . . . (54)

54a.   Terminal spike regularly pistillate at the summit, rarely more than 3 mm wide; lateral spikes rarely more than 3 mm wide; scales whitish or brownish-tinged, much shorter than the perigynia; plants densely tufted; culms up to 40 cm long; inflorescence up to 15 cm long; terminal spike mostly staminate, 4–8 mm long; pistillate spikes 2 or 3, drooping on slender peduncles. **Carex capillaris** L.; Fig. 99. In moist moss in open spruce woodland; frequent.

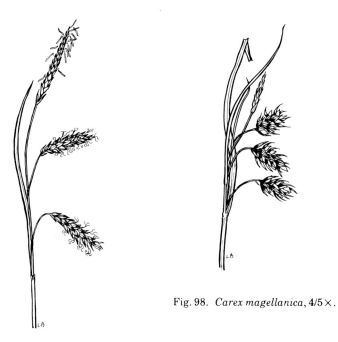

Fig. 98. *Carex magellanica*, 4/5×.

Fig. 97. *Carex limosa*, 4/5×.

Fig. 99. *Carex capillaris*, 1/2×.

54b.   Terminal spike regularly staminate throughout or
         sometimes with a few perigynia at the summit . . . (55)

55a.   Perigynia beakless or with a short essentially entire
         beak up to 0.5 mm long, less than one-quarter the
         length of the body; tufted plants, from short rhizomes;
         culms up to 80 cm long; inflorescence up to 15 cm long;
         terminal spike staminate, often hidden by pistillate
         spikes; pistillate spikes 1–2 cm long, with the lower
         ones long-peduncled; bracts leaf-like, exceeding the
         spikes. **Carex granularis** Muhl. Wet lake margins;
         rare.

55b.   Perigynia with the beak one-quarter the length of the
         body or equaling it . . . . (56)

56a.   Staminate spike sessile or short-peduncled; pistillate
         spikes subglobose to ellipsoid; culms 30 cm long,
         caespitose, characteristically yellowish green;
         inflorescence up to 5 cm long; staminate spike
         terminal, about 1 cm long, often almost hidden by the
         2–6 crowded pistillate spikes; perigynia 2.0–3.5 mm
         long, green or brownish green, about equaled by the
         yellowish brown scales. **Carex viridula** Michx.; Fig.
         100. Lakeshores and moist clearings; common around
         Lake Katherine, occasional elsewhere.

56b.   Staminate spike or spikes long-peduncled; pistillate
         spikes oblong-cylindrical or narrower . . . . (57)

57a.   Plant with slender horizontal yellowish brown
         rhizomes; pistillate spikes linear, loosely
         3–20-flowered; perigynia 3–4 mm long, light brown,
         beaked; scales brownish purple, smaller, loose, acute;
         culms up to 60 cm long, solitary or few together;
         inflorescence up to 15 cm long; terminal spike
         staminate, 1–2 cm long, peduncled; pistillate spikes
         1–3, short to long-peduncled, 10–25 mm long, spread-
         ing or drooping. **Carex vaginata** Tausch; Fig. 101.
         Moss in moist woods, clearings and shores; frequent.

57b.   Plants caespitose; pistillate spikes oblong-cylindrical,
         rather closely flowered; scales pale brown with
         whitish margins . . . . (58)

58a.   Pistillate spikes 1–5 cm long, 8–10 mm wide;
         staminate spikes 1–4; perigynia 5–6 mm long, with the
         body subglobose and contracted into a bidentate beak
         about as long as the body; scales shorter, somewhat
         acute or blunt; tufted plants from long, creeping
         rhizomes; culms up to 80 cm long; inflorescence up to

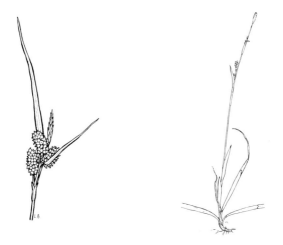

Fig. 100. *Carex viridula*, 3/5×.   Fig. 101. *Carex vaginata*, 1/4×.

20 cm long. **Carex sprengelii** Dewey. Scrub prairie, moist open woods, and clearings; frequent.

58b.   Pistillate spikes 4–15 mm long, 2.5–4.0 mm wide; scales acute or blunt; staminate spike usually solitary. See *Carex capillaris*.

Eleocharis   spike-rush

1a.   Tubercule confluent with the 3-angled achene, forming a beak; spikelets 3–6 mm long; plants in small tufts, from long, creeping rhizomes; culms up to 30 cm long. **Eleocharis pauciflora** (Lightf.) Link; Fig. 102. Shore of Katherine Lake; localized.

1b.   Tubercule distinct . . . . (2)

2a.   Achenes 3-angled; stigmas 3; spikelets flattened, 2–7 mm long, 3–15-flowered; plants forming dense mats, from slender rhizomes; slender tufted culms up to 15 cm long. **Eleocharis acicularis** (L.) R. & S.; Fig. 103. Mud flats and lakeshores; occasional.

2b.   Achenes lenticular or biconvex; stigmas 2 . . . . (3)

95

3a.　　Sterile basal scale 1, encircling the culm; spikelets 1.0–1.5 mm long, loosely few-flowered; achenes 1.0–1.4 mm wide; tubercule often as broad as high, 0.6–1.0 mm broad at the base; plants loosely tufted, from slender reddish rhizomes; culms slender, up to 50 cm long. ***Eleocharis uniglumis*** (Link) Schultes; Fig. 104. Mucky depressions; rare.

3b.　　Sterile basal scales 2 or 3; spikelets up to 2.5 cm long; culms up to 60 cm tall and 3 mm thick . . . . (4)

4a.　　Tubercule much longer than broad; plants from long, creeping reddish rhizomes. ***Eleocharis palustris*** (L.) R. & S.; Fig. 105. Swamps, lakeshores, and wet depressions; frequent.

4b.　　Tubercule as broad as or broader than long; similar to *E. palustris*. ***Eleocharis smallii*** Britt. Dried depression on shale on east slope; rare.

Eriophorum   cotton-grass

1a.　　Spikelets more than 1, on spreading or drooping peduncles; inflorescence subtended by one or more leafy bracts . . . . (2)

1b.　　Spikelets solitary, erect, not subtended by a leafy involucre . . . . (4)

2a.　　Involucral bract solitary; leaves 1.0–1.5 mm wide, channeled to the base; tufted slender culms up to 40 cm long, mostly without basal leaves; inflorescence of 2–5 spikelets. ***Eriophorum gracile*** Koch; Fig. 106. Quaking fen; localized.

2b.　　Involucral bracts 2 or more; leaves 1.5–8.0 mm wide, flat at least below the middle . . . . (3)

3a.　　Upper leaf sheaths dark-girdled at the summit; midrib of scales not extending to the tip; anthers 2.5–5 mm long; culms up to 60 cm long, mostly solitary, from short stout rhizomes; inflorescence consisting of 2–10, 1–2-cm-long divergent or drooping spikelets; bristles 2–5 cm long. ***Eriophorum angustifolium*** Honck.; Fig. 107. Moist spruce woodland and bogs; rare.

3b.　　Upper leaf sheaths not dark-girdled at the summit; midrib of scales extending to the tip; anthers 1.0–1.3 mm long; culms up to 70 cm long, in small tufts; inflorescence consisting of 3–15, 5–10-mm-long

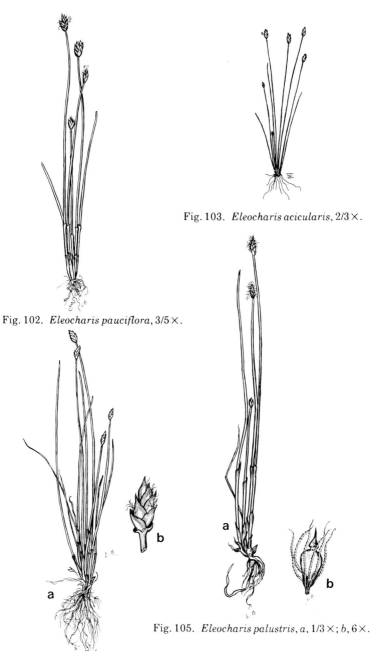

Fig. 103. *Eleocharis acicularis*, 2/3 ×.

Fig. 102. *Eleocharis pauciflora*, 3/5 ×.

Fig. 105. *Eleocharis palustris*, a, 1/3 ×; b, 6 ×.

Fig. 104. *Eleocharis uniglumis*, a, 2/5 ×; b, 2 ×.

97

divergent or drooping spikelets; bristles 1–2 cm long. ***Eriophorum viridi-carinatum*** (Engelm.) Fern.; Fig. 108. Fen; rare.

4a.   Plants with rhizomes; culms up to 35 cm long, solitary or few together; fruiting heads globose to obovoid, 2.5–4.0 cm long; scales with broad whitish margins; anthers 1.5–3.0 mm long. ***Eriophorum chamissonis*** C.A. Mey. Sedge-grass meadow; rare.

4b.   Plants densely tufted, lacking rhizomes; culms up to 60 cm long, stiff; head 1.0–1.5 cm long; bristles 2.0–2.5 cm long; scales lead gray to blackish, divergent or reflexed; anthers about 2.0 mm long. ***Eriophorum vaginatum*** L. ssp. ***spissum*** (Fern.) Hult. (*E. spissum* Fern.); Fig. 109. Spruce bogs; rare.

Scirpus   bulrush

1a.   Involucre none or merely the modified scale of the terminal solitary spike; culms round in cross section, wiry, smooth, up to 30 cm long, densely tufted, forming hard tussocks. ***Scirpus caespitosus*** L. ssp. ***austriacus*** (Pall.) Asch. & Graeb. (var. *callosus* Bigel.); Fig. 110. Open calcareous fen; rare.

1b.   Involucre consisting of 1 to many mostly leaf-like bracts . . . . (2)

2a.   Involucre consisting of a single, firm, erect bract, appearing to be a continuation of the culm; culms round, soft, easily compressed, naked or leafy only toward the base; spikelets mostly terminating, divergent rays of the inflorescence. ***Scirpus validus*** Vahl; Fig. 111. Shallow water bordering lakes and streams, marshes, and ditches; frequent.

2b.   Involucre consisting of 2 or more flat leaves; culms leafy . . . . (3)

3a.   Plants rhizomatous; culms up to 80 cm long; leaf sheaths reddish; leaves 4–15 mm wide, flat; inflorescence up to 20 cm long; branches 3–15 cm long, with the shorter ones ascending and the longer ones drooping; spikelets 3–6 mm long; bristles downwardly barbed, short. ***Scirpus microcarpus*** Pers. (*S. rubrotinctus* Fern.); Fig. 112. Beaver meadows, wet stream banks, and lakeshores; frequent.

Fig. 106. *Eriophorum gracile*, 2/3 ×.

Fig. 108. *Eriophorum viridi-carinatum*, 1/2 ×.

Fig. 107. *Eriophorum angustifolium*,
a, 2/5 ×; b, 2/5 ×.

Fig. 109. *Eriophorum vaginatum* ssp.
spissum, 2/5 ×.

3b.   Plants tufted, nonrhizomatous; leaves 2–5 mm wide; inflorescence up to 20 cm long; branches 3–10 cm long, ascending to drooping; spikelets 3–6 mm long; bristles smooth, longer than the scale, elongating at maturity. ***Scirpus cyperinus*** (L.) Kunth. Ditches and borders of ponds; rare.

17. ARACEAE   arum family

1a.   Leaves and spathes narrow and sword-like; spadix appearing lateral on the stem . . . . ***Acorus***
1b.   Leaves and spathes broad; spadix terminal . . . . ***Calla***

Acorus   sweetflag

Plants aromatic; rhizome thick, creeping; fruit becoming dry, but gelatinous inside, 1- to few-seeded. ***Acorus calamus*** L.; sweetflag; Fig. 113. Bordering streams; apparently rare, but perhaps overlooked.

Calla   water-arum

Low perennial, from long rhizomes rooting at the nodes; leaves cordate, long-stalked; inflorescence stalked; spike-like spadix backed by a white spathe; fruits red berries, few-seeded. ***Calla palustris*** L.; wild calla, water-arum; Fig. 114. Swamps and shallow water.

18. LEMNACEAE   duckweed family

1a.   Thallus with 1 or more rootlets . . . . (2)
1b.   Thallus without rootlets . . . . ***Wolffia***

2a.   Rootlet 1; thallus 1–5-nerved . . . . ***Lemna***
2b.   Rootlets 2 to several; thallus 4–15-nerved . . . . ***Spirodela***

Lemna   duckweed

1a.   Thallus 2–5 mm long, round or elliptical; flowers tiny and only rarely seen; reproduction mainly by tiny

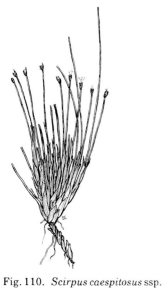

Fig. 111. *Scirpus validus*, 4/5 ×.

Fig. 110. *Scirpus caespitosus* ssp.
austriacus, 2/5 ×.

Fig. 112. *Scirpus microcarpus*, 1/3 ×.

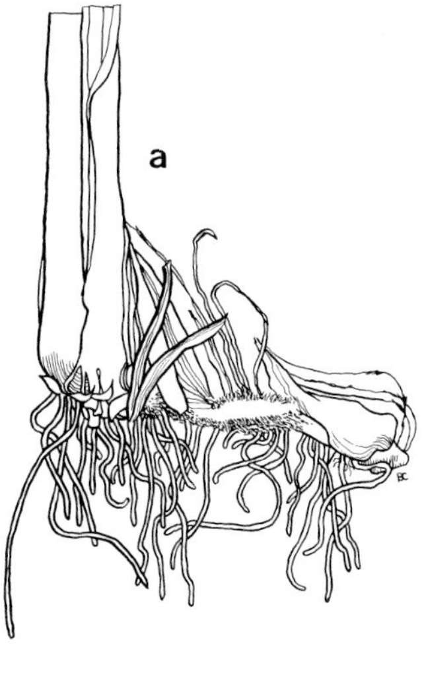

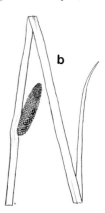

Fig. 113. *Acorus calamus, a,* 2/5 ×; *b,* 1/4 ×.

buds that form along the edge of the parent frond. **Lemna minor** L.; duckweed. Often forming a dense scum on the surface of stagnant water.

1b.   Thallus 6–10 mm long, 3-lobed, stalked; reproduction mainly asexual as *L. minor*. **Lemna trisulca** L.; star duckweed; Fig. 115. Plants forming submerged tangled mats, often among the stems of sedges and other plants in quiet streams, marshes, and beaver ponds.

Spirodela   water-flaxseed

Thallus 3–8 mm long, round-obovate, purple and somewhat convex below, dark green above. **Spirodela polyrhiza** (L.) Schleid.; water-flaxseed, larger duckweed. Floating singly on quiet water of streams and ponds, often with *Lemna minor*, or sometimes stranded.

Wolffia   watermeal

Thallus 0.7–1.5 mm long, globose to ellipsoid, light green; reproduction almost always by budding. **Wolffia columbiana** Karst.; watermeal. Forms a dense green cover often with *Lemna minor*, on the surface of some beaver ponds.

19. JUNCACEAE   rush family

1a.   Capsule with many small seeds; plants never hairy
.... *Juncus*
1b.   Capsule 3-seeded; leaves and young stems hairy
.... *Luzula*

Juncus   rush

1a.   Annual; stems tufted, erect or spreading; inflorescence diffuse, about half the height of the plant; flowers borne singly or in twos or threes. **Juncus bufonius** L.; toad rush; Fig. 116. Low moist ground; frequent.
1b.   Perennial .... (2)

2a.   Flowers reddish-brown, in a few dense, globose glomerules; stoloniferous, forming dense colonies; stem and leaves thin and wiry. **Juncus nodosus** L.;

Fig. 115. *Lemna trisulca*, 4/5×.

Fig. 114. *Calla palustris*, 1/5×.

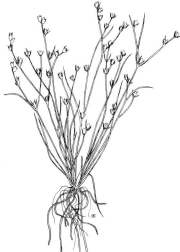

Fig. 116. *Juncus bufonius*, 2/5×.

knotted rush; Fig. 117. Moist low ground and lakeshores; frequent.

2b. Flowers in few-flowered glomerules . . . .(3)

3a. Leaves quill-like, hollow, with regularly spaced cross walls, 1 or 2 occurring on the erect culm. Culms close together along the rhizome; inflorescence on more or less divergent branches; glomerules 3–12-flowered, dense, less than hemispheric. ***Juncus alpinus*** Vill.; alpine rush; Fig. 118. Moist depressions and lakeshores; apparently rare.

3b. Leaves not hollow . . . . (4)

4a. Bract of inflorescence terete, appearing as a continuation of the stem . . . . (5)

4b. Bract of inflorescence flat or channeled; inflorescence appearing terminal or lateral . . . . (6)

5a. Perianth segments green; involucral leaf up to 20 cm long or longer; culms wiry, arising in a line along the elongate, cord-like rhizome; leaves reduced almost to bladeless sheaths. ***Juncus filiformis*** L.; Fig. 119. Bogs and lakeshores; apparently rare.

5b. Perianth segments purplish brown; involucral leaf to about 10 cm long; culms wiry, arising in a line along the elongate, cord-like rhizome; leaves reduced almost to bladeless sheaths. ***Juncus balticus*** Willd. var. ***littoralis*** Engelm.; Baltic rush; Fig. 120. Lakeshores, borders of sloughs, wet meadows, and sandy situations; frequent.

6a. Culms flattened, leafy, with at least one leaf borne at or above the middle; rhizome becoming slender and elongate; inflorescence usually overtopped by an elongate bract. ***Juncus compressus*** Jacq. Wet ground; introduced.

6b. Culms terete; fine leaves confined to the lower third of the culm; rhizome ascending; inflorescence overtopped by an elongate bract. ***Juncus dudleyi*** Wieg. Wet situations; frequent.

Luzula   wood-rush

1a. Flowers in pale brown to yellowish green glomerules; tufted plants up to 40 cm high; cauline leaves 3 or 4, about the same length as the basal leaves; leaves thickened at the tip and sparsely long-ciliate; inflorescence overtopped by a leafy bract. ***Luzula multiflora*** (Retz.) Lej.; field wood-rush. Open mixed woods and undulating prairie; apparently rare.

1b. Flowers single or sometimes 2 together at the ends of obvious peduncles in a subglobose umbel; cauline leaves short, much smaller than elongate basal leaves; leaves thickened at the tip and very long-ciliate; bract of the inflorescence shorter than, to about the same length as, the pedicels of the umbel. ***Luzula pilosa*** (L.)

Fig. 117. *Juncus nodosus*, 2/5×.

Fig. 118. *Juncus alpinus*, 3/5×.

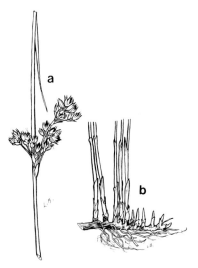

Fig. 119. *Juncus filiformis*, 2/5×.

Fig. 120. *Juncus balticus* var. *littoralis*, a, 3/5×; b, 2/5×.

Willd. var. **americana** R. & S. (*L. acuminata* of Am. auth.). Moist woodlands often hidden amongst low shrubs and thus perhaps overlooked.

20. LILIACEAE lily family

1a. Leaves basal or nearly so . . . . (2)
1b. Leaves alternate or in whorls on the stem . . . . (3)

2a. Flowers in umbels; scape from a coated bulb . . . . **Allium**
2b. Flowers in racemes; scapes from short or creeping rhizomes . . . . **Tofieldia**

3a. Plants climbing by tendrils . . . . **Smilax**
3b. Plants not climbing, without tendrils . . . . (4)

4a. Flowers solitary . . . . (5)
4b. Flowers in clusters . . . . (8)

5a. Leaves netted-veined, broad, in a single whorl of 3 . . . . **Trillium**
5b. Leaves parallel-veined, alternate or in several whorls . . . . (6)

6a. Flowers large and very showy . . . . **Lilium**
6b. Flowers 1–4.5 cm long . . . . (7)

7a. Flowers orange yellow or straw yellow; fruit a capsule . . . . **Uvularia**
7b. Flowers greenish; fruit a bright red or orange red berry . . . . **Disporum**

8a. Flowers large and very showy . . . . **Lilium**
8b. Flowers to 4.5 cm long . . . . (9)

9a. Flowers in umbels . . . . **Disporum**
9b. Flowers in racemes or panicles . . . . (10)

10a. Fruit a capsule; styles 3, separate . . . . **Zygadenus**
10b. Fruit a berry; styles sometimes cleft at the tip . . . (11)

11a. Perianth segments 4; leaves 2 or 3, broadest towards the cordate base . . . . **Maianthemum**
11b. Perianth segments 6; leaves 2 to several, tapering to the base . . . . **Smilacina**

Allium
onion

1a.   Leaves terete, hollow; flower pedicels shorter than the individual flowers; perianth parts 10–12 cm long, pink. ***Allium schoenoprasum*** L. var. ***sibiricum*** (L.) Hartm.; wild chives; Fig. 121. Shorelines; occasional.

1b.   Leaves flat; flower pedicels longer than the individual flowers; perianth parts up to 8 mm long, pink. ***Allium stellatum*** Fraser; wild onion. Prairie and parkland; occasional.

Disporum
fairybells

A branched herb up to 80 cm high; leaves alternate, ovate to oblong-lanceolate, cordate, subsessile; flowers 1–3 at the ends of the branches; perianth parts 8–14 mm long, creamy white; berry depressed-globose, densely papillose, orange red. ***Disporum trachycarpum*** (S. Wats.) B. & H.; fairybells. Moist woods and ravines.

Lilium
lily

Plants up to 60 cm high or higher, from whitish, scaly bulblets; leaves linear, in whorls (or the uppermost leaves whorled and the other leaves alternate in **L. *philadelphicum*** L. var. ***andinum*** (Nutt.) Ker); flowers 1–3(–5), very showy; perianth parts about 8 cm long, red or orange with black spots; fruit a capsule 3–5.5 cm long. ***Lilium philadelphicum*** L.; wood lily. Open woods, clearings, and scrub prairie; frequent.

Maianthemum
wild lily-of-the-valley

Low stoloniferous herbs; sterile leaves cordate, numerous; fertile stems fewer, up to 15 cm long; ovate leaves 2 or 3; raceme consisting of small white, sweetly fragrant flowers; berries pale red. ***Maianthemum canadense*** Desf. var. ***interius*** Fern.; wild lily-of-the-valley; Fig. 122. Rich moist woods; occasional.

Smilacina
false Solomon's-seal

1a.   Erect plants up to 50 cm high; leaves alternate; 6–12, folded and overlapping in youth, spreading and flattening in age; flowers small, white, in a spike-like raceme; berries green with black stripes. ***Smilacina stellata*** (L.) Desf.; star-flowered Solomon's-seal; Fig. 123. Thickets, woodlands, and meadows; frequent.

1b. Low erect plants up to 30 cm high; leaves alternate; 2–4; flowers small, white, in a spike-like raceme; berries bright red. ***Smilacina trifolia*** (L.) Desf.; three-leaved Solomon's-seal; Fig. 124. Bogs; localized.

Smilax   carrionflower

Climbing herbs with pairs of tendrils from the axils of the middle and upper leaves; leaves broadly cordate, net-veined; flowers small, greenish, in globular, long-pedunculate umbels; fruit a deep blue berry with a glaucous bloom. ***Smilax herbacea*** L. var. ***lasioneuron*** (Hook.) DC.; carrionflower. Rich woodland near east entrance; rare.

Tofieldia   asphodel

Tufted plants, from a short rhizome; leaves linear, basal, about the length of the scape; scape glandular, topped by a fascicle of whitish flowers; capsule stramineous or red. ***Tofieldia glutinosa*** (Michx.) Pers.; sticky asphodel; Fig. 125. Open calcareous fen; rare.

Trillium   wakerobin, trillium

Plants up to 40 cm high, growing from short rhizomes; leaves in a verticil of 3, rhombic-ovate; flowers on pedicels reflexed below the leaves; petals white; fruit a red berry. ***Trillium cernuum*** L.; nodding trillium; Fig. 126. Rich moist woodland; rare.

Zygadenus   camas

An onion-like plant up to 60 cm high, with long, pale green basal leaves and shorter stem leaves; flowers racemose or paniculate, yellow with a dark glandular patch towards the base; fruit an ovoid capsule. ***Zygadenus elegans*** Pursh; white camas; Fig. 127. Clearings and scrub prairie; frequent; poisonous.

Fig. 121. *Allium schoenoprasum* var. *sibiricum*, 2/5×.

Fig. 122. *Maianthemum canadense* var. *interius*, 1/4×.

Fig. 124. *Smilacina trifolia*, 2/5×.

Fig. 123. *Smilacina stellata*, 1/3×.

21. IRIDACEAE   iris family

Sisyrinchium   **blue-eyed-grass**

Tufted grass-like plants with 2-edged stems; flowers blue; petals about 1 cm long, mucronate; capsule orbicular. *Sisyrinchium montanum* Greene; blue-eyed-grass; Fig. 128. In sod in meadows and scrub prairie; frequent.

22. ORCHIDACEAE   orchid family

1a.	Flowers single or rarely 2 or 3	(2)
1b.	Flowers in a raceme	(3)

2a.	Leaf 1, basal	*Calypso*
2b.	Leaves 2 or more, cauline	*Cypripedium*

3a.	Leaves 2, opposite	*Listera*
3b.	Leaves alternate or all basal	(4)

4a.	Lower petal spurred	(5)
4b.	Lower petal not or sometimes obscurely spurred	(6)

5a.   Lip white, spotted with purple; sepals and petals roseate . . . . ***Orchis***
5b.   Flowers uniformly greenish or white . . . . . ***Habenaria***

6a.   Plants without chlorophyll; leaves reduced to scales . . . . ***Corallorhiza***
6b.   Plants with chlorophyll; leaves normal . . . . (7)

7a.   Leaves basal, ovate to obovate, frequently strongly reticulate . . . . ***Goodyera***
7b.   Leaves basal or cauline, or both . . . . (8)

8a.   Leaves basal and cauline, linear . . . . ***Spiranthes***
8b.   Leaves basal, ovate to obovate . . . . (9)

9a.   Leaves more than 2, frequently strongly reticulate; scape with bracts . . . . ***Goodyera***
9b.   Leaves 2; scape bractless . . . . ***Liparis***

Fig. 125. *Tofieldia glutinosa*, a, 1/3×; b, 2/5×.

Fig. 127. *Zygadenus elegans*, a, 1/4×; b, 4/5×.

Fig. 126. *Trillium cernuum*, 1/4×.

Fig. 128. *Sisyrinchium montanum*, 1/4×.

111

Calypso   Venus-slipper

Low plants, up to about 15 cm high; basal leaf single, round-ovate; flower single, showy pink; lip vaguely slipper-shaped. *Calypso bulbosa* (L.) Oakes; Venus-slipper; Fig. 129. In feathermoss under white spruce; rare.

Corallorhiza   coralroot

1a.   Lip 3-lobed or with a prominent lateral tooth on each margin .... (2)
1b.   Lip unlobed, madder purple, abruptly drooping; sepals and petals purplish, conspicuously purple-veined; capsules strongly reflexed. *Corallorhiza striata* Lindl.; striped coralroot. Rich woods; rare.

2a.   Plant greenish or yellowish green, up to 30 cm high; flowers yellowish green to slightly brown-tinged; lip notched on each side toward the base, with low basal lobes, white and unspotted or rarely dotted with red or purple; capsules greenish; flowering chiefly in spring. *Corallorhiza trifida* Chat.; early coralroot; Fig. 130. Rich woods; rare.
2b.   Plant madder purple or brownish, up to 50 cm high; flowers usually spotted with purple or red; lip auricled and with prolonged basal lobes; capsules brown or fulvous; flowering in summer. *Corallorhiza maculata* Raf.; spotted coralroot. Rich woods; rare.

Cypripedium   lady's-slipper

Erect leafy stems up to 40 cm long; leaves elliptical, more or less sheathing; flowers 1–2(–3), with an erect bract at the base; sepals and petals brownish and twisted; lip sac-like, yellow; var. *parviflorum* (Salisb.) Fern. has sepals 3–5 cm long and lip 20–35 mm long; var. *pubescens* (Willd.) Correll has sepals 5–8 cm long and lip 35–60 mm long. *Cypripedium calceolus* L.; yellow lady's-slipper; Fig. 131. Open woodland and clearings and ditches; becoming rare because of picking.

Goodyera   rattlesnake-plantain

Small herb; leaves in a rosette, ovate, usually white reticulate; scape, bracted, up to 30 cm long; raceme 1-sided; flowers small, white, glandular-downy. *Goodyera repens* (L.) R. Br.; dwarf rattlesnake-plantain; Fig. 132. In moist moss under spruce and balsam; rare.

Fig. 129. *Calypso bulbosa*, 2/5×.

Fig. 131. *Cypripedium calceolus*, 2/5×.

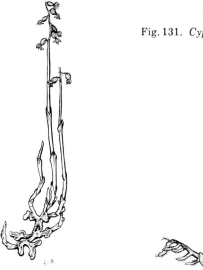

Fig. 130. *Corallorhiza trifida*, 2/5×.

Fig. 132. *Goodyera repens*, 1/2×.

113

Habenaria   bog orchid

1a.   Leaves basal .... (2)
1b.   Leaves cauline .... (3)

2a.   Leaf 1, obovate, ascending; scape bractless, up to 25 cm
       long; raceme loose; flowers greenish yellow; spur 5–7
       mm long, ***Habenaria obtusata*** (Pursh) Richardson
       (*Platanthera obtusata* (Pursh) Lindl.); blunt-leaf
       orchid; Fig. 133. Moist woods, bogs, and sometimes in
       scrub prairie; frequent.
2b.   Leaves 2, orbiculate to broadly elliptical, flat on the
       ground; scape bracted, up to 50 cm long; raceme lax,
       open; flowers greenish white; spur slenderly clavate,
       reflexed, 1.6–2.7 mm long. ***Habenaria orbiculata***
       (Pursh) Torr. (*Platanthera orbiculata* (Pursh) Lindl.);
       round-leaved orchid; Fig. 134. Rich moist woods; rare.

3a.   Lower bracts 1.5–6 times as long as the subtended
       flowers; stems stout, up to 60 cm long; leaves
       lanceolate or oblanceolate; raceme close; flowers
       greenish; spur 2–3 mm long; lip lingulate, with 2 short
       oblong or deltoid teeth and a small median tooth.
       Habenaria viridis (L.) R.Br. var. ***bracteata*** (Muhl.)
       Gray (*Coeloglossum bracteatum* (Muhl.) Parl.); frog
       orchid; Fig. 135. Moist woods, clearings and scrub
       prairie; occasional.
3b.   Bracts shorter .... (4)

4a.   Flowers white, spicy fragrant; lip abruptly widened at
       the base; stems slender, up to 60 cm long; leaves often
       narrow; raceme spike-like. ***Habenaria dilatata***
       (Pursh) Hook. (*Platanthera dilatata* (Pursh) Lindl.);
       leafy white orchid, bog-candle; Fig. 136. Open
       calcareous fen; rare.
4b.   Flowers greenish; lip gradually widened toward the
       base; stems small to stout, up to 60 cm high; leaves
       narrowly oblong to oblong-lanceolate; raceme
       cylindrical, open to dense. ***Habenaria hyperborea*** (L.)
       R. Br. (*Platanthera hyperborea* (L.) Lindl.); northern
       green orchid; Fig. 137. Bogs, moist woodland,
       clearings, and meadows; frequent.

Liparis   twayblade

       Stems up to 20 cm long, strongly ribbed; 2 leaves basal,
broadly lanceolate, sheathing; flowers yellowish green, about
5 mm wide, on short ascending pedicels. ***Liparis loeselii*** (L.)

Fig. 133. *Habenaria obtusata*, 1/2 ×.

Fig. 135. *Habenaria viridis* var. *bracteata*, 2/5 ×.

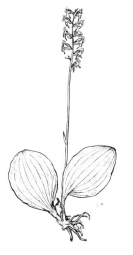

Fig. 134. *Habenaria orbiculata*, 1/5 ×.

Fig. 136. *Habenaria dilatata*, 1/4 ×.

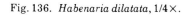

Rich.; twayblade. Feathermoss carpet at edge of marl fen; rare.

Listera   twayblade

Plants low, rather delicate, up to 20 cm high; leaves in a pair, broadly round-ovate, fixed about the middle; flowers racemose, green, tinged with purple. ***Listera cordata*** (L.) R. Br.; heart-leaved twayblade; Fig. 138. Moist moss in spruce woods; very rare.

Orchis   orchid

Plants low; leaf solitary, orbicular to elliptical; scape naked, up to 25 cm high; raceme bracted; flowers 2–9; sepals and upper petals roseate; lip white, spotted with purple. ***Orchis rotundifolia*** Banks (*Amerorchis rotundifolia* (Banks) Hult.); small round-leaved orchid; Fig. 139. Bogs and deep moss under spruce; localized.

Spiranthes   ladies'-tresses

1a.   Flowers white, spreading horizontally, conspicuously dispersed in a single twisted, vertical row; leaves mostly basal, short stalked, ovate or elliptical, usually withering before the flowers appear. ***Spiranthes lacera*** Raf.; slender ladies'-tresses. Open woods and bogs; very rare.
1b.   Flowers white, in 3 rows; spike conspicuously twisted; leaves both basal and cauline, lanceolate to linear, with those of the stem smaller. ***Spiranthes romanzoffiana*** Cham.; hooded ladies'-tresses; Fig. 140. Bogs; rare.

23. SALICACEAE   willow family

1a.   Trees; leaves ovate or deltoid; buds with overlapping scales; aments arching or pendulous; bracts mostly lacerate .... ***Populus***
1b.   Trees or shrubs; leaves linear to lanceolate; buds with a single scale; aments ascending or divergent; bracts entire or subentire .... ***Salix***

Fig. 137. *Habenaria hyperborea*, 2/5×.

Fig. 139. *Orchis rotundifolia*, 2/5×.

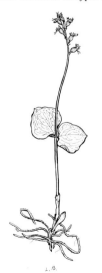

Fig. 138. *Listera cordata*, 2/5×.

Fig. 140. *Spiranthes romanzoffiana*, 2/5×.

117

Populus   poplar, aspen

1a. Leaf petioles terete; leaves ovate to ovate-lanceolate,
     acute to acuminate, cuneate to subcordate at the base,
     minutely crenulate to subentire, dark green and shiny
     above, much paler below with a conspicuous
     reticulation; tall trees up to 15 m high or higher; bark
     grayish white, becoming gray and furrowed in age;
     buds very resinous. **Populus balsamifera** L.; balsam
     poplar; Fig. 141. Low woods and shores and
     occasionally planted; common.
1b. Leaf petioles distinctly flattened . . . . (2)

2a. Leaves roundish in outline, abruptly short-tipped;
     margins finely crenate, slightly glaucous below,
     trembling with the slightest breeze; clonal tree, up to
     14 m high; bark pale grayish green or almost white
     because of a lichen coating. **Populus tremuloides**
     Michx.; aspen poplar, trembling aspen; Fig. 142.
     Usually well-drained situations; common.
2b. Leaves deltoid-cordate and caudate in outline;
     margins coarsely serrate and not as white below as *P.
     balsamifera*; trees up to 18 m high. **Populus ×jackii**
     Sarg. (*P. balsamifera × deltoides*). Low wet areas
     below the east slope and rare to the west.

Salix   willow

1a. Capsules (and ovaries) glabrous . . . . (2)
1b. Capsules (and ovaries) pubescent . . . . (11)

2a. Leaves entire or nearly so, linear-oblong to
     oblong-ovate, dull green and more or less glaucous
     above, glaucous below; catkins appearing at the same
     time as the leaves on leafy peduncles; low shrubs up to
     1.5 m high; twigs reddish yellow to brownish. **Salix
     pedicellaris** Pursh var. **hypoglauca** Fern.; Fig. 143.
     Bogs; localized.
2b. Leaves obviously toothed . . . . (3)

3a. Petioles bearing conspicuous glands on the upper
     surface near the base of the strongly reticulate-veined
     blades . . . . (4)
3b. Petioles glandless (or only weakly so in *S. fragilis*)
     . . . . (6)

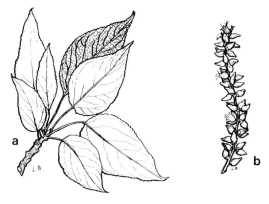

Fig. 141. *Populus balsamifera, a*, 1/4×; *b*, 2/5×.

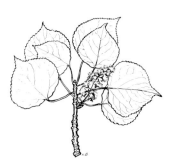

Fig. 142. *Populus tremuloides*, 1/4×.

Fig. 143. *Salix pedicellaris*, 1/3×.

4a.   Capsules tufted in whorls along the rachis of the ament; leaves ovate-lanceolate, acuminate, tapering to the base, pale green above, whitened below; catkins appearing the same time as the leaves and borne on leafy peduncles; shrubs up to 15 m high; twigs yellowish brown to reddish brown, drooping. **Salix amygdaloides** Anderss.; peach-leaved willow. Planted along docking area, Wasagaming.

4b.   Capsules spirally arranged . . . . (5)

5a.   Pistillate aments short, 1.5–3.0 cm long; capsules opening in late summer or autumn; leaves lanceolate

119

to elliptical-lanceolate, acute to short-acuminate; margins finely glandular-serrate, green and glossy above, lighter below; shrubs up to 4 m high or higher; twigs shiny yellowish brown. **Salix serissima** (Bailey) Fern.; autumn willow; Fig. 144. Low ground and wet shores; infrequent.

5b.   Pistillate aments longer, up to 7 cm long; capsules opening in late spring or early summer; leaves long-caudate, dark green and shining above, lighter below; shrubs or small trees up to 5 m high; twigs reddish brown. **Salix lucida** Muhl.; shining willow. Beaver meadows, marshes, and wet ditches; occasional.

6a.   Leaves sessile, or nearly so, linear to oblong-lanceolate, remotely sharp-toothed; catkins appearing at the same time as the leaves, on leafy peduncles; shrubs up to 4 m, from creeping rhizomes; twigs reddish. **Salix interior** Rowlee; sandbar willow; Fig. 145. Beaver meadows, marshes, lakes, and stream banks; frequent.

6b.   Leaves petioled, lanceolate to ovate . . . . (7)

7a.   Bracts of ament pale, falling before capsule matures; leaves lanceolate, glandular serrate, dark green above, glaucous below; catkins on leafy peduncles, appearing at the same time as the leaves; trees or shrubs; twigs yellow to reddish, breaking readily. **Salix fragilis** L. Low ground by water; rare.

7b.   Bracts of ament persistent until ripening of capsule
. . . . (8)

8a.   Bracts whitish, thin, and elongate; capsules 6–8 mm long on pedicels 2.5–3.5 mm long; leaves ovate to lanceolate-oblong, glandular-serrate to crenate, green above, glaucous and finely reticulate below, reddish when young, fragrant, especially when crushed; shrubs up to 4 m high; twigs greenish yellow when young, becoming reddish brown. **Salix pyrifolia** Anderss.; balsam willow; Fig. 146. Swampy areas; rare.

8b.   Bracts dark . . . . (9)

9a.   Catkins subsessile on naked peduncles, appearing before the leaves; leaves narrowly to broadly ovate, glandular-serrate to crenate, mostly subcordate at the base, green above, glaucous below, often reddish when young; shrubs up to 3 m high; twigs yellowish; stipules

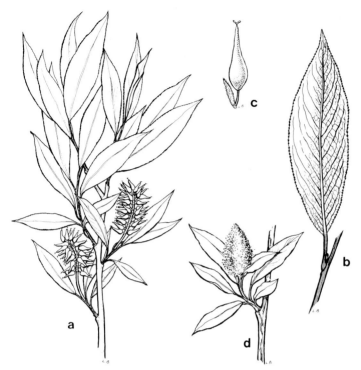

Fig. 144. *Salix serissima, a,* 1/2×; *b,* 7/8×; *c,* 2×; *d,* 2/3×.

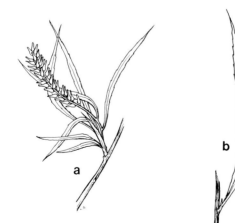

Fig. 145. *Salix interior, a,* 1/2×; *b,* 3/5×.

foliaceous. **Salix padophylla** Rydb. (*S. monticola* Bebb, *S. pseudomonticola* Ball); mountain willow; Fig. 147. Wet woods, depressions, and swamps; frequent.

9b. Catkins on short leafy-bracted peduncles, appearing at the same time or shortly before the leaves .... (10)

10a. Leaves elliptical to ovate, finely glandular-serrate to crenate, dark green and shining on both sides or only slightly paler beneath; shrubs up to 1 m high, but usually much shorter; twigs greenish to reddish brown. **Salix myrtillifolia** Anderss.; myrtle-leaved willow; Fig. 148. Low wet woods and bogs; apparently only occasional.

10b. Leaves lanceolate, acute to short acuminate, obtuse to somewhat cordate at the base, glaucous beneath; shrubs up to 5 m high; twigs yellow, becoming gray; stipules foliaceous. **Salix lutea** Nutt.; Fig. 149. Swamps, flooded depressions, lakeshores, and stream banks; occasional.

11a. Leaves distinctly toothed .... (12)
11b. Leaves entire or indistinctly toothed .... (15)

12a. Leaves mostly sessile or nearly so, linear to narrowly lanceolate, remotely sharp-toothed, green on both sides. See *Salix interior.*
12b. Leaves distinctly petioled, more or less glaucous beneath (except *S. maccalliana*, with leaves conspicuously reticulate-veined) .... (13)

13a. Catkins sessile, bractless; leaves elliptical to elliptical-oblanceolate, acute to short-acuminate; margins entire to undulate-crenate; shrubs or small trees up to 7 m high; twigs yellowish to reddish brown; stipules often present on vigorous shoots. **Salix discolor** Muhl.; pussy willow. Wet woods, swamps, lakeshores, and stream banks; common.
13b. Catkins on short leafy peduncles; stipules lacking .... (14)

14a. Leaves elliptical-lanceolate, green on both sides or only slightly paler on the strongly reticulate-veined lower surface; shrubs up to 3 m high; twigs reddish brown. **Salix maccalliana** Rowlee; Fig. 150. Marshes and beaver meadows; localized.
14b. Leaves narrowly to broadly lanceolate, acute at both ends, with the margins subentire to rather closely

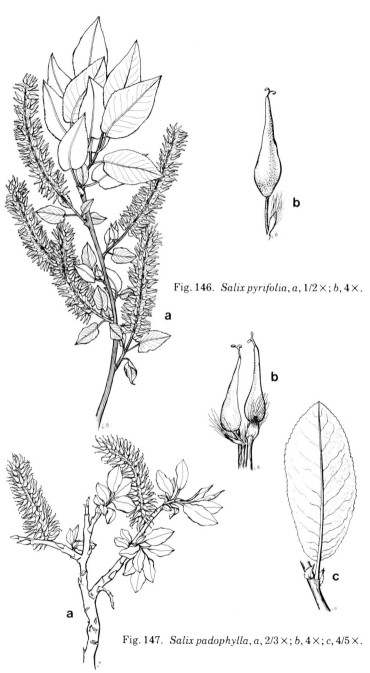

Fig. 146. *Salix pyrifolia*, *a*, 1/2×; *b*, 4×.

Fig. 147. *Salix padophylla*, *a*, 2/3×; *b*, 4×; *c*, 4/5×.

123

glandular-serrulate and more or less glaucous beneath but not strongly reticulate-veined; shrubs or small trees up to 7 m high; twigs yellowish becoming reddish. **Salix gracilis** Anderss. (*S. petiolaris* of Am. auth.); basket willow; Fig. 151. Swamps, flooded areas, and low wet second growth; common.

15a. Catkins sessile and leafless at the base . . . . (16)
15b. Catkins on leafy-bracted peduncles . . . . (18)

16a. Leaves densely silky-pubescent below, becoming glabrate in age, linear-lanceolate, acuminate; margins entire to obscurely serrate, revolute; shrubs or small trees up to 5 m high; twigs yellowish or greenish brown, brittle; catkins appearing just before or at the same time as the leaves. **Salix pellita** Anderss. Lakeshores; rare.
16b. Leaves dark green and glabrous above, glaucous and glabrous or only sparingly silky below . . . . (17)

17a. Capsules 7–12 mm long, clearly pedicellate. See *Salix discolor.*
17b. Capsules 6–7 mm long, subsessile; leaves elliptical to lanceolate, acute at both ends, glossy green with somewhat sunken veins above, glaucous below, subentire to glandular-serrate; shrubs up to 4 m high; twigs greenish brown, often pubescent at first but later glabrous; catkins appearing before the leaves. **Salix planifolia** Pursh; Fig. 152. Lakeshores, stream banks, and low meadows; common.

18a. Bracts greenish yellow with reddish tips; capsules 7–10 mm long on pedicels 3–6 mm long; styles obsolete or nearly so; leaves broadly oblanceolate, with the early ones villous or short sericeous when young and the later ones felty-tomentose below when young but becoming nearly glabrous; colonial bush or small tree up to 10 m high; twigs light brown, at first densely pubescent. **Salix bebbiana** Sarg.; Fig. 153. Swamps, ditches, and open woodland; common.
18b. Bracts brown; styles definite . . . . (19)

19a. Leaves dull white-tomentose below, lanceolate or narrower; margins revolute; shrubs up to 1 m high; twigs densely white-woolly. **Salix candida** Fluegge; hoary willow; Fig. 154. Bogs; localized.
19b. Leaves sparingly pubescent below or glabrous in age; margins not revolute. See *Salix gracilis.*

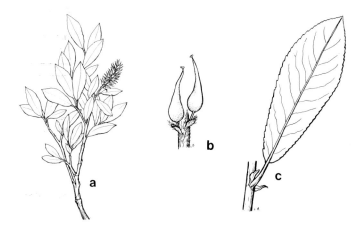

Fig. 148. *Salix myrtillifolia*, *a*, 1/3×; *b*, 2×; *c*, 7/8×.

Fig. 149. *Salix lutea*, *a*, 1/2×; *b*, 1 1/6×; *c*, 2×.

125

24. BETULACEAE   birch family

1a.   Nuts wingless, enclosed in a foliaceous involucre
. . . . **Corylus**
1b.   Nuts winged, without an involucre; both male and female flowers in scaly aments . . . . (2)

2a.   Female aments racemose, woody . . . . **Alnus**
2b.   Female aments single in leaf axils; bracts thin, deciduous . . . . **Betula**

Alnus   alder

1a.   Flowers produced at the same time as the leaves; leaves oval, finely and sharply serrate, glutinous; shrubs up to 3 m high; winter buds sessile; nutlets broadly winged. **Alnus crispa** (Ait.) Pursh; green alder, mountain alder; Fig. 155. Undergrowth in open woodland, clearings, and low areas; common.
1b.   Flowers produced after the leaves; leaves elliptical, rather coarsely doubly serrate, not glutinous; shrubs or small trees up to 5 m high; winter buds stipitate; nutlets essentially wingless. **Alnus incana** (L.) Moench ssp. **rugosa** (Du Roi) Clausen; speckled alder. Wet depressions, lakeshores, and stream banks; frequent.

Betula   birch

1a.   Trees up to 15 m high; bark white, papery; leaves ovate to rhomboid, up to 5 cm wide. **Betula papyrifera** Marsh.; white birch, paper birch, or canoe birch. Mixed woodland and clearings; common.
1b.   Shrubs up to 2 m high; leaves cuneate-obovate, up to 2 cm wide (wider in sucker growth). **Betula pumila** L. var. **glandulifera** Regel (*B. glandulosa* Michx. var. *glandulifera* (Regel) Gleason); swamp birch; Fig. 156. Swamps, open woodland, and clearings; common.

Corylus   hazelnut

1a.   Nuts barely covered by two distinct leafy bracts; shrubs up to 3 m high; twigs, petioles, and involucres more or less glandular-bristly; leaves broadly oval, up to 10 cm long. **Corylus americana** Walt.; American

Fig. 150. *Salix maccalliana*, a, 1/2×; b, 4/5×; c, 3×.

Fig. 151. *Salix gracilis*, 1/4×.

127

hazelnut. Thickets; apparently rare and perhaps restricted to the northwestern part of the park.
1b.   Nuts enclosed in united bracts that form a beak up to 3 cm long; shrubs similar but usually shorter than *C. americana*; not glandular bristly. **Corylus cornuta** Marsh.; beaked hazelnut. Dominant understory shrub.

25. FAGACEAE   beech family

Quercus   oak

Trees up to 15 m high on the eastern slopes, but much smaller in hilly scrub prairie situations; leaves deeply lobed, shiny green above, whitened and short woolly below; older branchlets sometimes with corky wings; acorn with a fringed cap. **Quercus macrocarpa** Michx.; bur oak. Well-drained slopes; localized; extensive stands on gravelly beach ridges adjacent to lower Ochre River and McCready Ski Road.

26. ULMACEAE   elm family

Ulmus   elm

Trees up to 15 m high or higher; leaves ovate-oblong to oval, abruptly pointed, doubly serrate, rough above; veins prominent; fruit a thin, flat, broadly winged, elliptic samara, maturing and falling in late spring. **Ulmus americana** L.; American elm or white elm. Rich lowlands, especially along streams throughout the park; frequent in the east.

27. CANNABACEAE   hemp family

Humulus   hop

Perennial climbing vine, trailing over shrubs and other vegetation; leaves opposite, toothed, palmately 3–7-lobed, with cordate bases; bracts subtending the inflorescences, often not lobed; male and female flowers on separate plants, with the male ones in showy loose panicles and the female ones in cone-like heads 2–5 cm long. **Humulus lupulus** L.; common hop. Waste places; occasional. Contact may cause dermatitis.

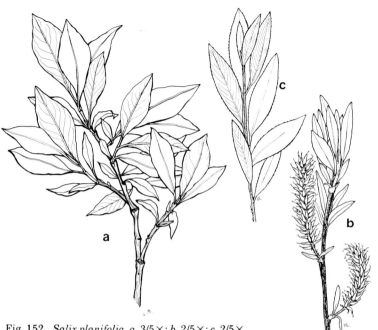

Fig. 152. *Salix planifolia*, a, 3/5×; b, 2/5×; c, 2/5×.

Fig. 153. *Salix bebbiana*, 1/2×.

28. URTICACEAE   nettle family

Urtica   nettle

Perennial plant from creeping rhizomes; squarish stems up to 1 m long; leaves opposite, ovate to lanceolate, serrate, with stinging hairs; flowers greenish, in the leaf axils. **Urtica dioica** L. ssp. **gracilis** (Ait.) Selander (*U. gracilis* Ait., *U. dioica* L. var. *procera* (Muhl.) Wedd.); stinging nettle; Fig. 157. Borders of sloughs, beaver dams, and moist areas; abundant but localized.

29. SANTALACEAE   sandalwood family

1a.   Flowers in terminal or subterminal clusters
.... ***Comandra***
1b.   Flowers axillary .... ***Geocaulon***

Comandra   comandra

Erect stems up to 30 cm long, usually several together, from a creeping rhizome; occasionally branched above; leaves alternate, linear or linear-lanceolate, sessile; flowers small, whitish to pinkish, several in terminal corymbs or panicles; fruit a dry nut surmounted by the free summit of the calyx. **Comandra umbellata** (L.) Nutt. (*C. pallida* A. DC., *C. richardsiana* Fern.); bastard toadflax. Scrub prairie and clearings; frequent.

Geocaulon

Erect stems up to 30 cm long; leaves alternate, elliptical to narrowly obovate, membranous, often purplish; flowers small, pedicellate; fruit a scarlet drupe. **Geocaulon lividum** (Richards.) Fern. (*Comandra livida* Richards.); northern comandra; Fig. 158. Rich moist woodland; rare.

30. POLYGONACEAE   buckwheat family

1a.   Stigmas capitate; sepals 5, often petaloid
.... ***Polygonum***
1b.   Stigmas tufted; sepals 6 .... ***Rumex***

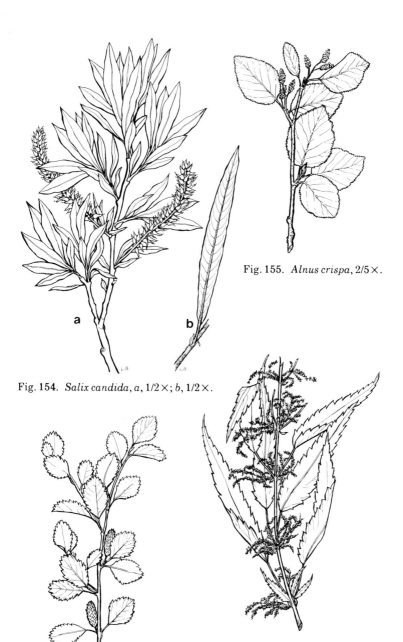

Fig. 155. *Alnus crispa*, 2/5×.

Fig. 154. *Salix candida*, *a*, 1/2×; *b*, 1/2×.

Fig. 157. *Urtica dioica* ssp. *gracilis*, 2/5×.

Fig. 156. *Betula pumila* var. *glandulifera*, 1/3×.

Polygonum   knotweed, smartweed

1a. Leaves hastate or cordate; plants twining over other vegetation .... (2)

1b. Leaves tapering at both ends; plants prostrate or upright, not twining .... (4)

2a. Sheaths with a ring of reflexed bristles at the base; leaves triangular-ovate, deeply cordate, pilose beneath; flowers in small racemes in the leaf axils or terminal; achenes lustrous. *Polygonum cilinode* Michx.; bindweed. Disturbed shale on eastern slope; rare.

2b. Sheaths beardless at the base; leaves glabrous, often scabrous on the veins beneath .... (3)

3a. Outer sepals merely keeled; flowers short-pediceled, in small racemes in the leaf axils or terminal; achenes dull black; basal lobes of leaves acute. *Polygonum convolvulus* L.; black bindweed, wild buckwheat. Introduced weed of waste ground; occasional.

3b. Outer sepals winged in fruit; wings thin and scarious; flowers in interrupted and bracteolate axillary racemes; achenes smooth and shining; basal lobes of leaves rounded. *Polygonum scandens* L.; climbing false buckwheat. Clearings and disturbed scrub areas; rare.

4a. Flowers in terminal, spike-like racemes; leaves not jointed to the petioles; sheaths mostly firm, rarely lacerate .... (5)

4b. Flowers in clusters in the leaf-axils or sometimes subtended by short bracts in a slender terminal spike; leaves jointed on very short petioles; sheaths finally 2-lobed or lacerate .... (10)

5a. Sheaths bristly-ciliate .... (6)

5b. Sheaths without cilia .... (8)

6a. Summit of sheath expanded into a horizontally divergent herbaceous flange or not; leaves oblong-lanceolate to linear-lanceolate; weak stems, from creeping blackish rhizomes; flowers pinkish red in terminal spikes 1–5 cm long. *Polygonum amphibium* L.; water smartweed; Fig. 159. Muddy shores and shallow water; occasional.

6b. Summit of sheath without a spreading flange .... (7)

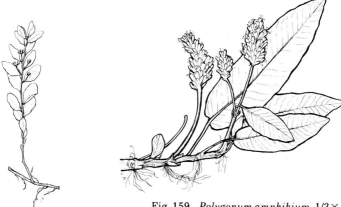

Fig. 159. *Polygonum amphibium*, 1/3×.

Fig. 158. *Geocaulon lividum*, 1/4×.

7a.   Mature calyx glandular-punctate; spikes slender, arching; achenes lustrous; stems up to 60 cm long, simple or branched, erect or ascending. **Polygonum hydropiper** L.; common water smartweed. Wet places.

7b.   Mature calyx not glandular-punctate; spikes dense, oblong cylindrical; stems up to 80 cm long, erect or ascending; leaves often purplish-blotched above. **Polygonum persicaria** L.; lady's-thumb. Waste ground; rare.

8a.   Peduncles glabrous or nearly so . . . . (9)

8b.   Peduncles pubescent or sparsely to copiously glandular, or both. See **Polygonum lapathifolium**.

9a.   Spikes thick-cylindrical or ovoid, 1–4 cm long; leaves floating or submersed; perennial. See *Polygonum amphibium*.

9b.   Spikes slenderly cylindrical, 1–8 cm long; leaves aerial, tapering to short petioles, glandular-dotted or pubescent below; simple to branched annual. **Polygonum lapathifolium** L. Shores and waste places; occasional.

10a.   Flowers solitary or in pairs, soon reflexed, mostly subtended by short bracts in a slender terminal spike; stem and branches sharply angled, erect. **Polygonum douglasii** Greene. Dry prairie openings; rare.

10b.   Flowers not reflexed, in axillary clusters; stem round, prostrate, or ascending . . . . (11)

11a.   Leaves crowded, elliptical to ovate; sheaths nonfibrous; achenes olivaceous. **Polygonum achoreum** Blake; waste places; rare.

11b.   Leaves scattered, linear to oblong or elliptical; sheaths fibrous; achenes dark brown. **Polygonum aviculare** L.; knotweed. Introduced weed along trails and roadsides and in waste places; frequent.

Rumex   dock

1a.   Wings of fruit produced into a few elongated acicular lobes; annual usually diffusely branched plant, up to 60 cm high; leaves pale green, lanceolate, with the lower ones truncate to cordate at the base; inflorescence dense, golden brown. **Rumex maritimus** L. var. **fueginus** (Phil.) Dusen; golden dock; Fig. 160. Stream banks and lakeshores; localized.

1b.   Wings of fruit entire to merely erose or denticulate (2)

2a.   Pedicels with a thickened articulation . . . . (3)

2b.   Pedicels not articulated . . . . (4)

3a.   Valves of the fruit rather cordate, about 4 mm wide, without grains or with only 1 small grain; erect perennial up to 1 m high; leaves lanceolate, up to 25 cm long, tapering to the petiole; inflorescence dense. **Rumex fennicus** Murb.; field dock. Introduced in waste places; rare.

3b.   Valves usually with large grains, each at least one-fifth the width of the valves; stems up to 1 m long, ascending to decumbent at the base, producing branches or leaf tufts in the axils of the leaves; leaves linear-lanceolate, pale green. **Rumex triangulivalvis** (Danser) Rech. f. (*R. mexicanus* Meisn., *R. salicifolius* Weinm.).; Fig. 161. Waste areas and low ground by lakes and streams; occasional.

4a.   Grains of the fruiting calyx none or only 1 on one of the valves; stems up to 1 m long or longer; leaves broadly oblong or lanceolate, often cordate at the base; inflorescence dense and conspicuous. **Rumex occidentalis** S. Wats.; western dock; Fig. 162. Wet places; occasional.

4b.   Grains of the fruiting calyx 3, one to each valve; stems up to 1.5 m long; leaves lanceolate, acute, or rounded at the base; inflorescence dense and conspicuous. **Rumex orbiculatus** Gray; water dock. Wet places, swamps, and shallow water; rare.

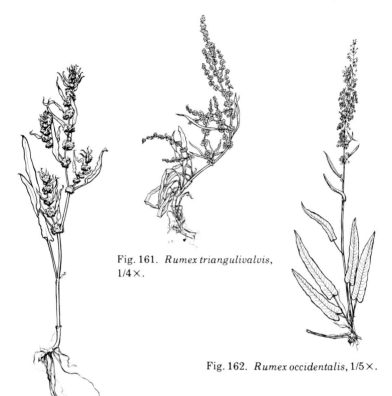

Fig. 161. *Rumex triangulivalvis*, 1/4×.

Fig. 162. *Rumex occidentalis*, 1/5×.

Fig. 160. *Rumex maritimus* var. *fueginus*, 1/2×.

31. CHENOPODIACEAE   goosefoot family

1a.	Fruit hidden between a pair of bracts ...... **Atriplex**
1b.	Fruit not hidden .... (2)
2a.	Calyx much reduced and not surrounding the fruit .... **Monolepis**
2b.	Fruit surrounded by the marcescent calyx .... (3)
3a.	Flowers unisexual, with the staminate ones borne in a conspicuously differentiated terminal spike ... **Axyris**
3b.	Flowers all perfect or some of them pistillate .... **Chenopodium**

135

Atriplex   orache

Erect annual up to 1 m high or higher; stems angular; leaves opposite or subopposite, succulent, lanceolate to linear-lanceolate, rarely ovate to oblong, often with a pair of out-pointing obtuse lobes; margins irregularly broad-toothed or entire; flowers loosely arranged in interrupted short glomerules on short to long stalks in the axils of the upper leaves; bracteoles thick, green, blackening at maturity. **Atriplex subspicata** (Nutt.) Rydb. Disturbed situations; rare.

Axyris

Erect bushy annual up to 60 cm high; leaves lanceolate, pale green; staminate flowers in terminal naked spikes of glomerules; pistillate flowers solitary, axillary. **Axyris amaranthoides** L.; Russian pigweed. Introduced weed; occasional.

Chenopodium   goosefoot, pigweed

1a. Fruit in large strawberry-like glomerules, partly axillary and partly in terminal leafy racemes; plants up to 40 cm high; leaves triangular-hastate, coarsely dentate, pale green. **Chenopodium capitatum** (L.) Aschers.; strawberry-blite; Fig. 163. Disturbed situations; rare.

1b. Fruit not fleshy or very slightly fleshy and the inflorescence less congested . . . . (2)

2a. Leaves mostly 1-nerved, narrow and entire, grayish-mealy at least below; tall herbs up to 60 cm high or higher; stems somewhat mealy, striate, or longitudinally grooved with alternate yellow and green lines. **Chenopodium leptophyllum** Nutt. Scrub prairie; rare.

2b. Leaves broader, with 1 or more nerves . . . . (3)

3a. Plants glabrous and green . . . . (4)
3b. Plants more or less mealy-puberulent . . . . (5)

4a. Seeds horizontal; testa smooth, unpatterned; plants up to 1 m high, branched; leaves bright green, thin, acute; base cordate to truncate, sinuate with 1–5 coarse acute teeth; inflorescence axillary and terminal, open. **Chenopodium gigantospermum** Aellen (*C. hybridum* L. var. *gigantospermum* (Aellen) Rouleau);

Fig. 163. *Chenopodium capitatum*, 1/3×.

maple-leaved goosefoot; Fig. 164. Disturbed situations; rare.

4b.   Seeds vertical (rarely horizontal); testa reticulate-punctate; plants up to 80 cm high, branched; leaves cuneate, triangular to rhomboid, fleshy, with conspicuous lateral teeth; inflorescence of glomerules in many compact, oblong cymes forming dense panicles or spikes. **Chenopodium rubrum** L.; coast-blite, red goosefoot. Borders of marshes and streams and in waste places; apparently rare.

5a.   Leaves 1-nerved, up to 4 cm long; margins undulate-dentate; teeth and tips obtuse; plants up to 40 cm high, erect to prostrate; inflorescence in small glomerules in axillary or terminal spikes. **Chenopodium glaucum** L.; oak-leaved goosefoot. Creek banks and waste places; rare.

5b.   Leaves 3-nerved .... (6)

6a.   Primary leaves thin, mainly linear, usually five times longer than wide, with 1 or 2 lobes and sometimes with teeth on the margin; inflorescence more or less leafy, forming a compact panicle. **Chenopodium pratericola** Rydb. Waste places; rare.

6b.   Primary leaves deltoid, rhombic, oblong or ovate, three times longer than wide .... (7)

7a.   Pericarp alveolate-reticulate or reticulate; stems up to 1 m long or longer, branched; leaves rather thick, rhomboid, ovate to lanceolate, entire to dentate; apex acute; inflorescence consisting of rounded glomerules in long spikes. **Chenopodium berlandieri** Moq. ssp. **zschackei** (Murr.) Zobel. Waste places; occasional.

7b.   Pericarp smooth or mottled .... (8)

8a.   Seeds circular in outline; pericarp nonadherent; stems up to 1 m long or longer, erect or ascending, simple to much branched; leaves ovate-lanceolate, varying to rhombic-lanceolate, sinuous-dentate to entire; inflorescence consisting of flowers clustered in elongate spikes of contiguous glomerules. **Chenopodium album** L.; lamb's-quarters, pigweed; Fig. 165. Introduced weed of waste places; frequent.

8b.   Seeds oval in outline; pericarp strongly adherent; stems up to 1 m long, erect or ascending, branched; leaves broadly oblong to ovate-lanceolate, cuneate at the base, shallowly serrate, greenish to reddish; inflorescence sparsely leafy; glomerules in terminal spikes. **Chenopodium strictum** Roth var. **glaucophyllum** (Aellen) Wahl. Disturbed waste areas; rare.

Monolepis   **povertyweed**

Branched prostrate herb up to 50 cm wide; stems fleshy, reddish; leaves narrow, hastate, passing gradually into foliaceous bracts; inflorescence in small axillary clusters. **Monolepis nuttalliana** (R. & S.) Greene; povertyweed. Waste ground; occasional.

32. AMARANTHACEAE   amaranth family

Amaranthus   **pigweed**

1a.   Prostrate, forming mats up to 60 cm wide; stems reddish, fleshy; leaves spatulate, broadest above the

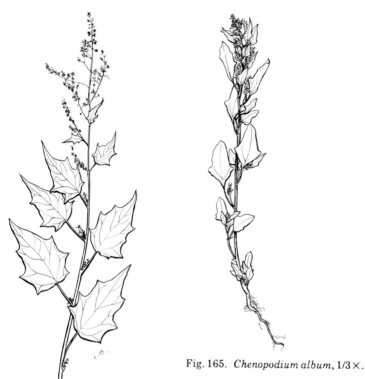

Fig. 165. *Chenopodium album*, 1/3×.

Fig. 164. *Chenopodium gigantospermum*, 1/3×.

middle, dark shiny green; flowers in the leaf axils. **Amaranthus graecizans** L. (*A. blitoides* S. Wats.); prostrate amaranth. Waste places; rare.

1b.   Stems erect, up to 1 m high, rough, angular, and somewhat hairy; leaves ovate, rough; inflorescence in dense spikes in the upper leaf axils and in a terminal spike at the summit, harsh and rough. **Amaranthus retroflexus** L.; red-root pigweed. Waste places; frequent.

33. PORTULACACEAE   purslane family

Portulaca   **purslane**

Fleshy annual forming mats up to 40 cm wide; leaves alternate, spatulate or obovate, dark shiny green; flowers yellow, borne singly in the leaf axils, but only open in bright sunshine; fruit a pointed capsule containing many minute seeds. **Portulaca oleracea** L.; purslane. Introduced garden weed.

34. CARYOPHYLLACEAE   pink family

1a.   Calyx with sepals distinct or united only at the base; petals without claws . . . . (2)
1b.   Calyx with sepals united into a tube; petals with basal claws . . . . (5)

2a.   Petals entire or merely notched at the apex . . . . . . (3)
2b.   Petals deeply cleft; capsules opening by twice as many valves or teeth as there are styles . . . . (4)

3a.   Capsules dehiscing by 6 teeth . . . . **Arenaria**
3b.   Capsules dehiscing by 3 teeth . . . . **Minuartia**

4a.   Styles 3; capsules short, ovate or oblong, opening by 6 valves . . . . **Stellaria**
4b.   Styles 5, opposite the sepals; capsules long, cylindrical, often bent or curved near the summit, opening by 10 teeth at the summit . . . . **Cerastium**

5a.   Calyx subtended by an involucre of long-tipped bracts . . . . **Dianthus**
5b.   Calyx naked at base . . . . (6)

6a.   Styles 5, alternate with the petals . . . . **Lychnis**
6b.   Styles 2 or 3 . . . . (7)

7a.   Styles 3; capsule 3-valved or 6-valved; calyx 10-nerved . . . . **Silene**
7b.   Styles 2; capsule 4-valved; calyx 5-nerved . . . . (8)

8a.   Leaves lance-acuminate; petals white, very small, about equaling the calyx . . . . **Gypsophila**
8b.   Leaves ovate-lanceolate; petals pale red, much longer than the calyx . . . . **Saponaria**

Cerastium   chickweed

1a.   Perennial matted plant, rooting at the nodes; leaves linear and narrowly lanceolate, short pubescent, not glandular; commonly with conspicuous axillary fascicles of short, sterile shoots that root when detached; inflorescence few- to many-flowered, on slender, elongated peduncles; petals much longer than the sepals. **Cerastium arvense** L.; field chickweed; Fig. 166. Scrub prairie and disturbed situations; frequent.

1b. Annual with erect stems; leaves linear-lanceolate; glandular-hirsute; inflorescence terminal, open cymose; petals equaling or shorter than the sepals. **Cerastium nutans** Raf. Wet situations by streams and ponds; occasional.

Dianthus   pink

Low trailing perennial with elongate basal offshoots; flowering stems up to 25 cm long; leaves linear-lanceolate, ciliate-margined and ciliate-keeled; flowers 1 to few, mostly long-stalked; petals pink, dentate. **Dianthus deltoides** L.; maiden-pink. Garden escape; rare.

Gypsophila   baby's-breath

Much-branched glaucous perennial up to 1 m high; leaves lance-acuminate, attenuate at the base; inflorescence a large panicle with corymbiform branches. **Gypsophila paniculata** L.; baby's-breath. Waste area at Whitewater Lake; localized.

Lychnis   campion

Leafy perennial up to 1.5 m high; leaves membranous, ovate below, lanceolate above, rounded or cordate at the base; flowers in dense terminal heads, each about 2.5 cm across; petals 2-cleft, scarlet. **Lychnis chalcedonica** L.; Maltese-cross, scarlet lychnis. Garden escape; rare.

Minuartia   sandwort

Loosely tufted branched annual or short-lived perennial; wiry stems up to 30 cm long; leaves linear, fresh green, 1-nerved, in fascicles from the nodes; inflorescence a few-flowered cyme; petals lacking or shorter than the strongly 3-nerved acuminate sepals. **Minuartia dawsonensis** (Britt.) Mattf. (*Arenaria dawsonensis* Britt); Fig. 167. Usually dry open and sometimes disturbed slopes; occasional.

Moehringia   sandwort

Stems up to 15 cm long or longer, simple or branched; leaves in 2–5 pairs, sessile, narrowly elliptical; flowers solitary or several in few-flowered cymes, occurring on very slender 2-bracted peduncles; petals white, longer than the sepals. **Moehringia lateriflora** (L.) Fenzl (*Arenaria lateriflora* L.); grove sandwort; Fig. 168. Open mixed woods; occasional.

Saponaria   soapwort

Annual with stems up to 70 cm long; leaves grayish green, ovate lanceolate, clasping at the base; flowers in loose corymbose cymes; petals white or pinkish, with an appendage at the top of the claw. **Saponaria officinalis** L.; bouncingbet. Garden escape; rare.

Silene   catchfly, campion

1a.   Calyx glabrous, much inflated; stems up to 60 cm long; leaves lanceolate, smooth; flowers in loose, open panicles; petals white, 2-cleft; **Silene vulgaris** (Moench) Garcke (*S. cucubalus* Wibel); bladder campion. Introduced weed of waste places.
1b.   Calyx glandular-pubescent, inflated or not ...... (2)

2a.   Leaves linear to narrowly oblanceolate; stems up to 40 cm long, viscid-puberulent; flowers on appressed-erect pedicels; calyx not inflated, tightly enclosing the capsule; petals white or purplish, included or barely exserted. **Silene drummondii** Hook. (*Lychnis drummondii* (Hook.) S. Wats., *L. pudica* Boivin). Open pineland; rare.
2b.   Leaves broader .... (3)

3a.   Flowers dioecious, fragrant; calyces of staminate flowers ellipsoid, of pistillate flowers ovoid and inflated at maturity; stems up to 1 m high, loosely forking; leaves oval to lance-oblong. **Silene alba** (Mill.) E.H.L. Krause (*Lychnis alba* Mill.); white cockle. Introduced weed of waste areas.
3b.   Flowers perfect, fragrant; calyx cylindrical, becoming inflated-ovoid; stems up to 1 m long, viscid-villous; upper leaves lanceolate; lower leaves ovate-lanceolate; basal leaves spatulate; inflorescence a small open cyme. **Silene noctiflora** L.; night-flowering catchfly. Introduced weed of waste areas.

Stellaria   chickweed, starwort

1a.   Flowers in the axils of green leaves or bracts ..... (2)

Fig. 166. *Cerastium arvense*, 2/5×.

Fig. 168. *Moehringia lateriflora*, 2/3×.

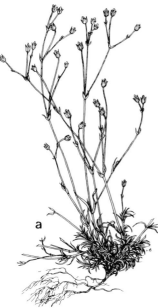

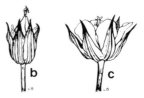

Fig. 167. *Minuartia dawsonensis*, *a*, 1/2×; *b*, 3×; *c*, 3×.

1b.   Inflorescence bracteolate; bracts membranous or membranous-margined .... (4)

2a.   Leaves elliptical to ovate or obovate, at least the lower ones long-petioled; stems weak, trailing, and matted; inflorescence diffuse to well defined, leafy to bracteolate. **Stellaria media** (L.) Cyrill.; common chickweed; Fig. 169. Introduced weed in gardens and around buildings.

2b.   Leaves sessile .... (3)

3a.   Petals about as long as or up to a little longer than the sepals; stems freely branching, matted; leaves linear oblong, somewhat fleshy; flowers in leafy cymes or terminal and solitary. **Stellaria crassifolia** Ehrh.; Fig. 170. Wet places; rare but perhaps overlooked.

3b.   Petals much shorter than the sepals or lacking; stems weak and trailing; leaves linear-lanceolate; flowers in a single terminal cyme. **Stellaria calycantha** (Ledeb.) Bong.; Fig. 171. Damp places; rare.

4a.   Pedicels ascending to erect; central flowers on more stiffly erect pedicels; stems forming tangled carpets; leaves green to glaucous, somewhat boat-shaped, narrowly lanceolate to linear, broadest near the base, gradually tapering to a sharp point. **Stellaria longipes** Goldie; Fig. 172. Grassy clearings and open woods; frequent.

4b.   Inflorescence more open, with some of the pedicels, especially those of the central flowers, spreading to deflexed; stems weak, forming tangled masses; leaves narrow, linear to linear-lanceolate. **Stellaria longifolia** Muhl.; Fig. 173. Swamps and wet places; occasional.

35. CERATOPHYLLACEAE   hornwort family

Ceratophyllaceae   **hornwort**

Stems long and branching; leaves verticillate, dichotomously divided into filiform segments; segments remotely serrulate; flowers monoecious, axillary, short-pediceled, inconspicuous. **Ceratophyllum demersum** L.; hornwort; Fig. 174. Completely immersed aquatic; forms dense masses in quiet water.

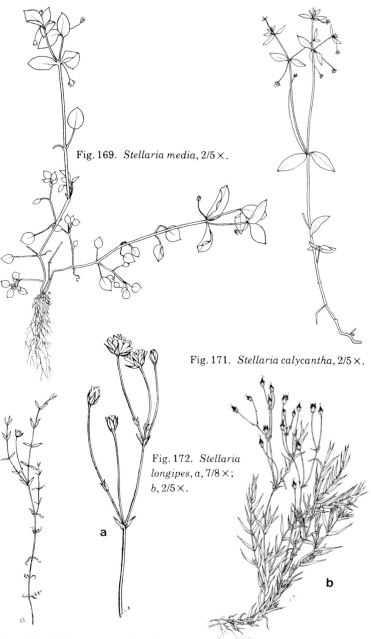

Fig. 169. *Stellaria media*, 2/5×.

Fig. 171. *Stellaria calycantha*, 2/5×.

Fig. 172. *Stellaria longipes*, a, 7/8×; b, 2/5×.

a

b

Fig. 170. *Stellaria crassifolia*, 1/2×.

145

36. NYMPHAEACEAE   water-lily family

Nuphar   **pond-lily**

1a.   Floating leaves elliptical-ovate, less than 10 cm long, deeply cordate at the base; flowers 1.5–2.0 cm wide; sepals yellow, 1.0–1.8 cm long; stigmatic disc red. ***Nuphar microphyllum*** (Pers.) Fern.; small pond-lily. Ponds and lakes; rare.

1b.   Floating leaves broadly ovate; blades up to 35 cm long; sinus closed and narrow; flowers 4–5 cm wide; sepals yellow, 2.0–3.5 cm long, often reddish towards the base; stigmatic disc green.   ***Nuphar variegatum*** Engelm.; yellow pond-lily; Fig. 175. Ponds and lakes; frequent.

37. RANUNCULACEAE   crowfoot family

1a.   Flowers spurred . . . . (2)
1b.   Flowers without spurs, regular . . . . (3)

2a.   Spur at base of petals; flowers regular, terminating the branches . . . . ***Aquilegia***
2b.   Spur at base of upper sepal, enclosing spurs of 2 upper petals; flowers irregular, in terminal racemes . . . . ***Delphinium***

3a.   Fruit red or white, berry-like . . . . ***Actaea***
3b.   Fruit not berry-like . . . . (4)

4a.   Fruit consisting of follicles (pods splitting down one side) . . . . (5)
4b.   Fruit consisting of numerous achenes on a rounded or elongated central axis . . . . (6)

5a.   Leaves basal, evergreen; leaflets 3, lustrous and sharply toothed, cuneate-obovate . . . . ***Coptis***
5b.   Leaves simple, orbicular to reniform, at least some of them borne on the stem . . . . ***Caltha***

6a.   Stem leaves opposite or whorled . . . . (7)
6b.   Stem leaves alternate . . . . (8)

7a.   Styles much elongating in fruit . . . . ***Pulsatilla***
7b.   Styles remaining short in fruit, not plumose . . . . ***Anemone***

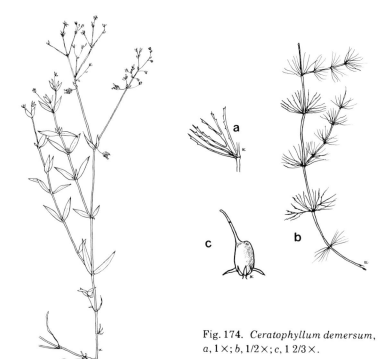

Fig. 174. *Ceratophyllum demersum*, a, 1×; b, 1/2×; c, 1 2/3×.

Fig. 173. *Stellaria longifolia*, 2/5×.

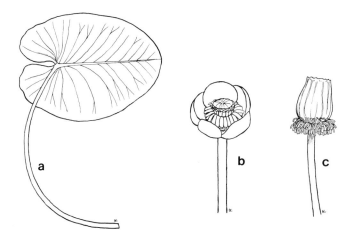

Fig. 175. *Nuphar variegatum*, a, 1/5×; b, 2/5×; c, 2/5×.

147

8a.   Leaves 3–4-ternately compound .... *Thalictrum*
8b.   Leaves simple, toothed (rarely entire) to deeply lobed or finely dissected .... *Ranunculus*

Actaea   baneberry

Stems up to 100 cm long; leaves large, compound; leaflets coarsely toothed; inflorescence a raceme at the end of the stem; sepals small, deciduous as the flower opens; fruit berry-like, red (f. *rubra*) or white (f. *neglecta* (Gilman) Robins.). *Actaea rubra* (Ait.) Willd.; red baneberry. Rich woodlands; frequent. Berries are reputed to be poisonous.

Anemone   anemone

1a.   Achenes nearly glabrous or densely short-hirsute, not woolly .... (2)
1b.   Achenes densely long-woolly, forming dense woolly heads .... (3)

2a.   Achenes glabrous or nearly so in a globular head; beak about as long as the body; stems up to 30 cm long, with several 5–7-parted, toothed basal leaves, a whorl of sessile involucral leaves, and a single white flower. *Anemone canadensis* L.; Canada anemone; Fig. 176. Clearings, road banks, and hollows; frequent.
2b.   Achenes densely short-hirsute; beak shorter than the body; stems up to 20 cm, with an involucre of 3 long-petioled deeply cut leaves; basal leaves 3–5-foliate; flowers solitary on a long pedicel. *Anemone quinquefolia* L.; wood anemone. Moist open woodland; occasional.

3a.   Leaves dissected into numerous linear or narrowly lanceolate acuminate lobes; stems up to 60 cm long, purplish toward the base; stem leaves short-petioled; flowers white, pink, or deep purple; fruiting head globose; achenes long-lanate. *Anemone multifida* Poir.; cut-leaved anemone; Fig. 177. Clearings, scrub prairie, and disturbed situations; occasional.
3b.   Leaves with 3–5 broad-lobed oblanceolate or obovate divisions .... (4)

4a.   Fruiting heads narrowly cylindrical; stems up to 50 cm long; leaves generally in a verticil of 5–7, trifoliate; leaflets 3-lobed; lobes coarsely few-toothed. *Anemone*

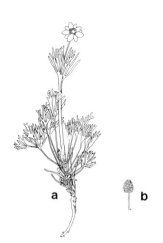

Fig. 177. *Anemone multifida, a,* 1/4×; *b,* 1/4×.

Fig. 176. *Anemone canadensis,* 1/4×.

cylindrica A. Gray; thimbleweed. Clearings and scrub prairie; occasional.

4b. Fruiting heads ovoid to thick cylindrical; stems up to 80 cm long; leaves generally in a verticil of 3, somewhat larger than in *A. cylindrica*; lobes serrate along the outer edge. **Anemone virginiana** L. (*A. riparia* Fern.); thimbleweed. Clearings, scrub prairie, and disturbed situations; occasional.

Aquilegia   columbine

1a. Petals blue or purple; spurs strongly hooked, longer than the blade of the petals; stems up to 50 cm long; basal leaves twice ternate; stem leaves smaller. **Aquilegia brevistyla** Hook.; blue columbine; Fig. 178. Wooded slope; rare.

1b. Petals scarlet or bright red, yellow within; spurs straight; tips merely oblique, shorter than the blade of the petals; stems up to 80 cm long or longer; plants similar to *A. brevistyla*, but more robust and the leaflets larger. **Aquilegia canadensis** L.; wild columbine. Clearings and borders of woodland; frequent.

Caltha   marsh-marigold

Coarse plants; stems hollow, erect or decumbent; leaves roundish to open-reniform, dentate; basal leaves with long stalks; upper leaves stalkless; flowers with bright yellow sepals; fruit a group of follicles. ***Caltha palustris*** L.; marsh-marigold; Fig. 179. Wet places; occasional.

Coptis   goldthread

Stems up to 15 cm long; leaves all basal, evergreen, long-stalked, each with 3 leaflets; flowers single; sepals 5–7, white. ***Coptis trifolia*** (L.) Salisb.; goldthread; Fig. 180. Rich woodland; rare.

Delphinium   larkspur

Stems up to 1.5 m long; leaves alternate, petioled, much divided or lobed; flowers racemose, perfect, irregular; upper sepal extended at the base into a spur. ***Delphinium glaucum*** S. Wats.; larkspur. Base of wooded bank along lakeshore and occasionally persisting after cultivation.

Pulsatilla   prairie crocus

Stems up to 30 cm long; basal leaves petioled, ternate; leaflets much divided into linear segments; stem leaves similar, 3, sessile; flowers large, appearing early in the spring; sepals bluish; styles elongating and plumose. ***Pulsatilla ludoviciana*** (Nutt.) Heller (*Anemone patens* L. var. *wolfgangiana* (Bess.) Koch); prairie crocus. Fig. 181. Well-drained slopes; rare.

Ranunculus   crowfoot, buttercup

1a.   Petals white (yellow at the base); fully submersed plant with the flowers either floating on the water or extending above it; leaves alternate, finely dissected into stiff filiform segments. ***Ranunculus aquatilis*** L. var. ***subrigidus*** (Drew) Breitung (*R. circinatus* Sibth. var. *subrigidus* (Drew) Benson); white water crowfoot; Fig. 182. Shallow lakes and ponds; occasional.

1b.   Petals yellow; plants aquatic or terrestrial . . . . . . (2)

2a.   Plants submersed or creeping on mud, rooting at the nodes; underwater leaves divided into narrow, flat

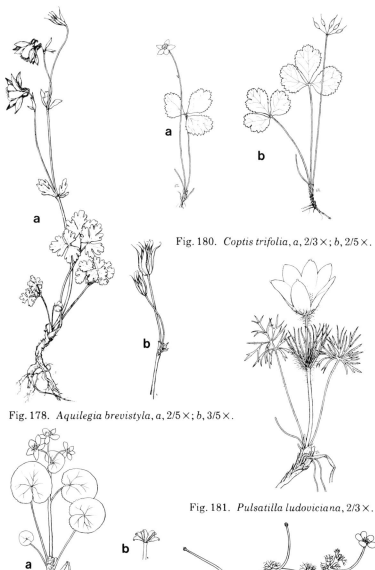

Fig. 180. *Coptis trifolia, a*, 2/3 ×; *b*, 2/5 ×.

Fig. 178. *Aquilegia brevistyla, a*, 2/5 ×; *b*, 3/5 ×.

Fig. 181. *Pulsatilla ludoviciana*, 2/3 ×.

Fig. 179. *Caltha palustris, a*, 1/8 ×; *b*, 1/4 ×.

Fig. 182. *Ranunculus aquatilis* var. *subrigidus*, 2/3 ×.

151

segments; floating leaves with wider segments. **Ranunculus gmelinii** DC.; yellow water crowfoot; Fig. 183. Shallow water of ponds and often stranded on mud; occasional.

2b. Plants terrestrial .... (3)

3a. Stems creeping, half-buried in mosses, up to 15 cm long; leaves alternate on the rhizome, trifid; segments trifid to lobed; cauline leaf smaller or absent; sepals 3; achenes with a hooked beak. **Ranunculus lapponicus** L.; Fig. 184. Mossy woods and bogs; rare.

3b. Stems mostly erect; sepals 5 .... (4)

4a. Plants tufted, with filiform repent stolons rooting at the nodes and giving rise to new plants; leaves long-petioled, roundish or reniform, crenate or dentate; stems up to 15 cm long, 1- to several-flowered; fruiting heads globose-ovoid to cylindrical. **Ranunculus cymbalaria** Pursh; seaside crowfoot; Fig. 185. Wet, chiefly alkaline places; localized.

4b. Plants without stolons .... (5)

5a. Achenes beakless or only minutely beaked ...... (6)
5b. Achenes distinctly beaked .... (8)

6a. Basal and mid-cauline leaves long-petioled, deeply palmately lobed or divided, thick; stems up to 60 cm long, hollow; flowers small and numerous. **Ranunculus sceleratus** L.; cursed crowfoot, celery-leaved buttercup; Fig. 186. Wet ground and shallow water; frequent.

6b. Basal leaves usually merely crenately toothed or shallowly lobed; cauline leaves sessile or short-petioled, mostly deeply divided, distinctly different from the basal ones .... (7)

7a. Plants glabrous or nearly so, up to 50 cm long; flowers 6–10 mm wide; petals usually shorter than the reflexed sepals. **Ranunculus abortivus** L.; kidney-leaved buttercup; Fig. 187. Damp clearings and shores; occasional.

7b. Plants pilose-hirsute, up to 40 cm high; flowers 10–20 mm wide; petals much exceeding the sepals. **Ranunculus rhomboideus** Goldie; prairie buttercup; Fig. 188. Scrub prairie; frequent.

8a. Petals 5–15 mm long, distinctly longer than the sepals; stems up to 80 cm long, hairy; basal leaves deeply

Fig. 183. *Ranunculus gmelinii*, 2/3×.

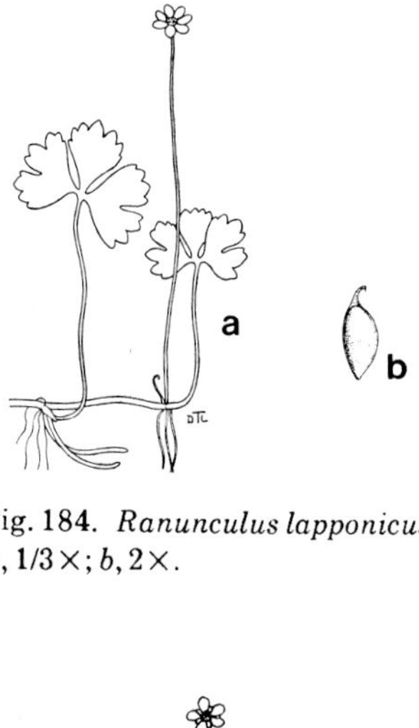

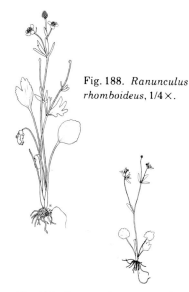

Fig. 186. *Ranunculus sceleratus*, 1/4×.

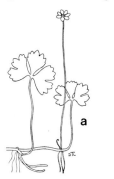

a

b

Fig. 184. *Ranunculus lapponicus*, a, 1/3×; b, 2×.

Fig. 188. *Ranunculus rhomboideus*, 1/4×.

Fig. 187. *Ranunculus abortivus*, 1/4×.

Fig. 185. *Ranunculus cymbalaria*, 2/3×.

153

palmately divided; sessile divisions deeply cleft. **Ranunculus acris** L.; common buttercup; waste places; rare.

8b. Petals 2–6 mm long, shorter than to barely exceeding the sepals .... (9)

9a. Flowers 10–15 mm wide; achenes 2.7–3.3 mm long in a globose or ovoid head; stems up to 60 cm long, hirsute, often decumbent; leaves in broad divisions with the segments usually stalked. **Ranunculus macounii** Britt.; Fig. 189. Low, moist places; frequent.

9b. Flowers 6–8 mm wide; achenes 1.8–2.7 mm long, in a thick cylindrical head; stems up to 60 cm long, hirsute; leaves similar to *R. macounii*. **Ranunculus pensylvanicus** L. f.; bristly crowfoot; Fig. 190. Wet places; rare.

Thalictrum   meadow-rue

1a. Plants up to 1 m high or higher; leaves ternately divided into many leaflets; leaflets coriaceous, pubescent at least on the lower surface, not glandular, mostly tri-lobed; flowers in a large terminal panicle. **Thalictrum dasycarpum** Fisch. & Lall.; purple meadow-rue. Moist thickets and clearings; occasional.

1b. Plants up to 50 cm high; leaflets smaller, flabellately tri-lobed; lobes 3-toothed, glabrous or minutely glandular-puberulent. **Thalictrum venulosum** Trel.; Fig. 191. Moist thickets and clearings; frequent.

38. FUMARIACEAE   fumitory family

Corydalis   corydalis

Stems lax and sometimes prostrate, diffusely branched; leaves much dissected, glaucous; flowers golden yellow, racemose; fruit pod-like, spreading or pendulous. **Corydalis aurea** Willd.; golden corydalis; Fig. 192. Usually in disturbed situations in lighter soils; occasional.

39. CRUCIFERAE   mustard family

1a. Capsules 1 or 2(–3) times as long as broad .... (2)
1b. Capsules (3 or)4 to many times as long as broad ... (5)

Fig. 189. *Ranunculus macounii*,
a, 1/6×; b, 1/4×.

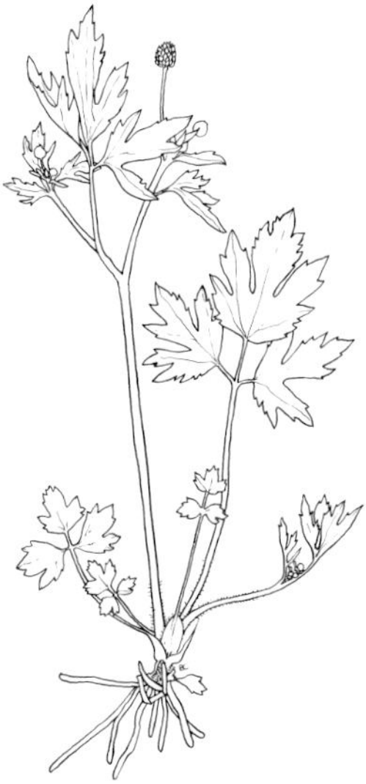

Fig. 190. *Ranunculus pensylvanicus*, 1/4×.

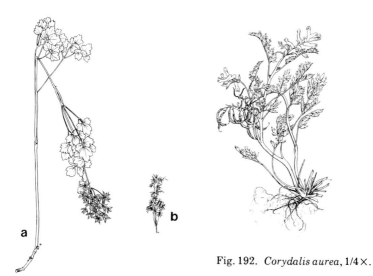

Fig. 192. *Corydalis aurea*, 1/4×.

Fig. 191. *Thalictrum venulosum*, a, 1/4×; b, 1/4×.

2a.   Capsule more or less flattened . . . . (3)
2b.   Capsule scarcely if at all flattened . . . . **Rorippa**

3a.   Capsule triangular . . . . **Capsella**
3b.   Capsule orbicular or obovate . . . . (4)

4a.   Capsule 1.0–1.8 cm long, broadly winged . . . . **Thlaspi**
4b.   Capsule 2.0–3.5 mm long, wingless or narrowly winged at the summit . . . . **Lepidium**

5a.   Stems 15–30 cm long or longer, mostly leafy towards the base . . . . **Draba**
5b.   Stems longer, leafy . . . . (6)

6a.   Stem leaves sessile, with auriculate or sagittate clasping bases . . . . (7)
6b.   Stem leaves sessile or petioled, not clasping . . . . . . (9)

7a.   Flowering stems arising from a definite rosette of basal leaves . . . . **Arabis**
7b.   Flowering stems without definite rosettes . . . . (8)

8a.   Pods with a large flat or angled beak, often containing a seed . . . . **Sinapis**
8b.   Pods with a slender beak; beak round or conical, without seeds . . . . **Brassica**

9a.   Stem leaves entire or toothed, only the basal ones sometimes lobed . . . . (10)
9b.   Stem leaves (or most of them) deeply cleft, pinnatifid to tripinnatifid or pinnate . . . . (12)

10a.   Stems hairy; hairs closely appressed straight, 2-pronged (malpighian), attached near their middle; flowers yellow . . . . **Erysimum**
10b.   Stems glabrous or pubescent (but hairs not malpighian); flowers purple, mauve or white . . . . (11)

11a.   Flowering stems without basal rosettes . . . . **Hesperis**
11b.   Flowering stems arising from a definite rosette of basal leaves . . . . **Arabis**

12a.   Racemes with lower pedicels subtended by leafy bracts . . . . **Erucastrum**
12b.   Racemes bractless . . . . (13)
13a.   Capsules flattened, linear; petals white or roseate; leaves pinnate . . . . **Cardamine**

13b.   Capsules round or 4-angled in cross section; flowers
        yellow .... (14)

14a.   Pubescence of leaves and stem stellate or forked;
        leaves finely divided .... **Descurainea**
14b.   Pubescence of leaves and stem simple or wanting;
        leaves pinnatifid to pinnate .... **Sisymbrium**

Arabis   rock cress

1a.   Pods 3–6 cm long, spreading to descending; stems
        purplish, up to 50 cm long; leaves sessile,
        narrow-lanceolate; basal leaves stellate-pubescent;
        flowers mauve, in a terminal raceme.   **Arabis
        divaricarpa** A. Nels.; Fig. 193.   Disturbed situations;
        rare.
1b.   Pods ascending to erect, mostly straight .... (2)

2a.   Pedicels and pods closely appressed to the rachis and
        parallel to one another .... (3)
2b.   Pedicels and pods more or less divergent.   See *Arabis
        divaricarpa.*

3a.   Pods about 1 mm wide, cylindrical to flattened ... (4)
3b.   Pods 1.5–3.0 mm wide, strongly flattened; stems up to
        1 m long; basal leaves and base of stem with
        malpighian hairs. **Arabis drummondii** A. Gray; Fig.
        194. Scrub prairie and disturbed situations; rare.

4a.   Stems up to 60 cm long, glabrous except near the base;
        rosette leaves slightly hairy; flowers yellowish white.
        Arabis glabra (L.) Bernh.; tower mustard.   Scrub
        prairie and disturbed situations; rare.
4b.   Stems up to 60 cm long, hairy about up to the middle;
        leaves coarsely hairy; flowers white or greenish white.
        Arabis hirsuta (L.) Scop. spp. **pycnocarpa** (Hopkins)
        Hult.; Fig. 195. Banks and clearings; occasional.

Brassica   mustard

    Stems up to 80 cm long, somewhat branched; leaves
thickish; lower leaves lyrate with a large terminal lobe; upper
leaves smaller and usually entire, deeply cordate, clasping;
flowers yellow. **Brassica campestris** L.; bird-rape. Introduced
weed around buildings and along roadsides; occasional.

Capsella   shepherd's-purse

Stems branched, up to 50 cm long; basal leaves pinnatifid; stem leaves lanceolate, auriculate-clasping; flowers racemose, small, white; pods an inverted triangle, flat. ***Capsella bursa-pastoris*** (L.) Medic.; shepherd's-purse. Introduced weed of waste places; frequent.

Cardamine   bitter cress

Stems up to 50 cm long; more or less branched; leaves deeply pinnately lobed; terminal lobe largest; flowers small, white; pods 1–3 cm long, divergent. ***Cardamine pensylvanica*** Muhl.; bitter cress; Fig. 196. Wet places; rare.

Descurainia   tansy mustard

1a.   Pod oblanceolate, 5–10 mm long, not more than twice as long as the strongly ascending pedicel; calyx 1.0–1.5 mm long; petals yellow; stems up to 90 cm long; leaves pinnate to doubly pinnate, grayish pubescent. ***Descurainia richardsonii*** (Sweet) O.E. Schulz; gray tansy mustard; Fig. 197. Disturbed situations; occasional.

1b.   Pod linear, 1–3 cm long, at least twice as long as the spreading-ascending pedicel; calyx 2.0–2.5 mm long; petals yellow; stems up to 96 cm long; leaves finely bipinnate to tripinnate. ***Descurainia sophia*** (L.) Webb.; flixweed. Introduced weed of disturbed places; frequent.

Draba   whitlow-grass

Leaves broadly lanceolate, soft pubescent on both sides; flowers small; petals pale yellow; raceme elongated and lax in fruit; capsules narrowly clavate, about 10 mm long, on slender spreading pedicels twice as long as the body. ***Draba nemorosa*** L.; Fig. 198. Disturbed situations on scrub prairie; rare.

Erucastrum   dog mustard

Erect plants, up to 50 cm long; leaves oblong, deeply pinnatifid; at least the lower flowers of the raceme leafy bracted; petals pale yellow; pods 3–5 cm long, tipped with a short slender beak. ***Erucastrum gallicum*** (Willd.) O.E. Schulz; dog mustard. Introduced weed of waste places; occasional.

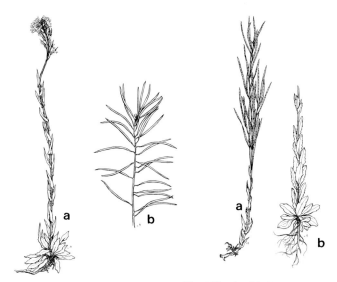

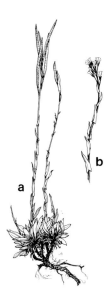

Fig. 193. *Arabis divaricarpa*, a, 1/4×; b, 1/8×.

Fig. 195. *Arabis hirsuta* ssp. *pycnocarpa*, a, 1/4×; b, 1/4×.

Fig. 196. *Cardamine pensylvanica*, 1/4×.

Fig. 194. *Arabis drummondii*, a, 1/5×; b, 1/4×.

159

Erysimum treacle mustard

1a.   Petals 3–4 mm long, pale yellow; pods less than 2 cm long; peduncles slender, ascending, about one-half as long as the straight pod; stems up to 50 cm long, branched or unbranched; leaves lanceolate or oblong-lanceolate, dark green, hairy; hairs mostly short-stalked, 3- to 4-forked. ***Erysimum cheiranthoides*** L.; wormseed mustard; Fig. 199. Weedy; occasional.

1b.   Petals about 10 mm long; pods 3–5 cm long; peduncles stout, curved, about one-fifth as long as the pod; stems up to 50 cm long, simple or branched mainly above; leaves narrow, with mostly malpighian hairs. ***Erysimum inconspicuum*** (Wats.) MacM.; small-flowered prairie-rocket; Fig. 200. Waste places; rare.

Hesperis rocket

Stems up to 100 cm long, simple to branched above; leaves oblong to ovate-lanceolate, dentate, pubescent on both sides; flowers large, 3–5 cm wide, fragrant; petals purple; pods 5–10 cm long, linear, ascending to spreading. ***Hesperis matronalis*** L.; dame's-rocket. Escaped from cultivation; rare.

Lepidium pepper-grass

Stems up to 60 cm long, much branched; stem leaves lanceolate or with a few coarse teeth; basal leaves incised; flowers minute; petals rudimentary or absent; pods on short stocks, very numerous. ***Lepidium densiflorum*** Schrad.; common pepper-grass. Weed of roadsides and waste places; frequent.

Rorippa yellow cress

Stems up to 60 cm long, branched or unbranched; leaves pinnatifid; terminal segment much larger than the lateral; flowers yellow; pods fat, oblong, about as long as the pedicels. ***Rorippa islandica*** (Oeder) Borbas; marsh yellow cress; Fig. 201. Wet places; frequent.

Sinapis charlock

Stems up to 80 cm long; lower leaves pinnatifid at the base, with a large terminal lobe; flowers yellow; pod with a beak

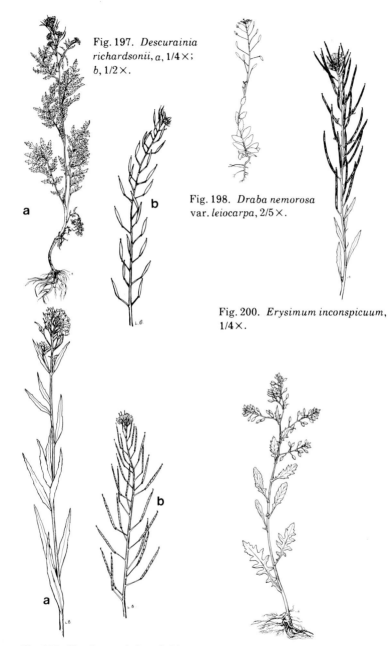

Fig. 197. *Descurainia richardsonii*, *a*, 1/4×; *b*, 1/2×.

Fig. 198. *Draba nemorosa* var. *leiocarpa*, 2/5×.

Fig. 200. *Erysimum inconspicuum*, 1/4×.

Fig. 199. *Erysimum cheiranthoides*, *a*, 2/5×; *b*, 2/5×.

Fig. 201. *Rorippa islandica*, 1/6×.

about half as long as the body. **Sinapis arvensis** L. (*Brassica kaber* (DC.) Wheeler var. *pinnatifida* (Stokes) Wheeler); charlock. Introduced weed of waste places; rare.

Sisymbrium

Stems up to 100 cm long, branched; leaves pale green; basal leaves pinnatifid; stem leaves pinnate to entire; flowers pale yellow; pods numerous, linear, 5–10 cm long, on short peduncles about as thick as the pods. **Sisymbrium altissimum** L.; tumbling mustard. Introduced weed; rare.

Thlaspi   pennycress

Stems up to 40 cm long, usually branched from the base; basal leaves oblanceolate, deciduous; stem leaves oblong to lanceolate, sinuate, clasping; flowers white, tiny; raceme elongating in fruit; pods ovate, flat, with a wide wing and a deep terminal notch. **Thlaspi arvense** L.; pennycress, stinkweed. Introduced weed of waste places; frequent.

40. DROSERACEAE   sundew family

Drosera   sundew

1a.   Leaves basal, spatulate or oblanceolate, with glandular sticky hairs that entrap insects and with the petiole much longer than the blade; flowers small, regular, in a 1-sided raceme; fruiting stems elongating to 15 cm or longer. **Drosera anglica** Huds.; Fig. 202. Open calcareous fen; localized and rare.

1b.   Similar, but the leaf blades almost round, broader than long. **Drosera rotundifolia** L.; round-leaved sundew; Fig. 203. Bogs and fens; localized.

41. SAXIFRAGACEAE   saxifrage family

1a.   Shrubs; fruit a berry . . . . **Ribes**
1b.   Herbs; fruit a capsule . . . . (2)
2a.   Petals lacking; flowers golden yellow . . . .
      . . . . **Chrysosplenium**
2b.   Petals present . . . . (3)

Fig. 202. *Drosera anglica*, 1/4×.   Fig. 203. *Drosera rotundifolia*, 2×.

3a. Flowers solitary .... ***Parnassia***
3b. Flowers racemose or panicled .... (4)

4a. Flowers irregular; calyx tube strongly oblique at the summit and swollen on one side at the base .... ***Heuchera***
4b. Flowers regular .... ***Mitella***

Chrysosplenium   **golden saxifrage**

Stems erect up to 15 cm long; leaves alternate, reniform, crenate; sepals yellowish green, with the outer ones somewhat wider than the inner ones; stamens 5–8, inserted on a conspicuous disc. ***Chrysosplenium alternifolium*** L. var. ***ioense*** (Rydb.) Boivin. (*C. ioense* Rydb.); golden saxifrage. In moss of damp woods; rare.

Heuchera   **alumroot**

Flowering stems up to 50 cm long, terminating in a panicle of purplish-petaled flowers; leaves basal, rounded-cordate;

teeth broadly ovate. ***Heuchera richardsonii*** R. Br.; alumroot; Fig. 204. Clearings and scrub prairie; frequent.

Mitella   miterwort

Flowering stems up to 25 cm long, terminated by a raceme of flowers; sepals 4, greenish; petals 5, finely pinnatifid, greenish white. ***Mitella nuda*** L.; bishop's-cap, miterwort; Fig. 205. Moist woodland; frequent.

Parnassia   grass-of-Parnassus

1a.   Petals 3 or more times longer than the sepals; staminodia 3–5 per cluster; stems up to 40 cm long, leafless or with a sessile leaf near the base; rosette leaves broadly ovate, petioled. ***Parnassia glauca*** Raf.; open calcareous fen; rare.

1b.   Petals 1.5 times longer than the sepals; staminodia 9–17 per cluster; stems up to 30 cm long, with a solitary sessile leaf borne near the middle; rosette leaves cordate at the base. ***Parnassia palustris*** L. var. ***neogaea*** Fern. (*P. multiseta* of Scoggan); Fig. 206. Moist clearings; occasional.

Ribes   currant

1a.   Flowers in elongate racemes; pedicels jointed below the ovary . . . . (2)
1b.   Flowers solitary or in clusters of 2–4; pedicels not jointed below the ovary . . . . (6)

2a.   Leaves resinous-dotted below; stems unarmed; ovary and fruit smooth . . . . (3)
2b.   Leaves not resinous-dotted . . . . (4)

3a.   Calyx white, 4–5 mm long; free segments exceeding the length of the tube; racemes erect or ascending; bracts soon deciduous; fruit black; erect shrubs up to 1.5 m high; leaves 3- or occasionally 5-lobed, pungent when crushed. ***Ribes hudsonianum*** Richards.; northern black currant; Fig. 207. Moist thickets; rare.

Fig. 204. *Heuchera richardsonii*, 1/4×.

Fig. 206. *Parnassia palustris* var. *neogaea*, 2/5×.

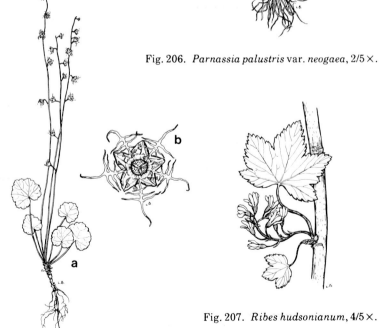

Fig. 207. *Ribes hudsonianum*, 4/5×.

Fig. 205. *Mitella nuda*, *a*, 2/5×; *b*, 3×.

165

3b.   Calyx yellow and whitish; lobes and tube subequal; racemes drooping; bracts elongate, persistent; fruit black; erect shrub up to 2 m high; leaves 3–5-lobed. **Ribes americanum** Mill. (*R. floridum* L'Her.); wild black currant. Moist thickets; occasional.

4a.   Stems up to 2 m high, armed with bristles and prickles; leaves deeply 5–7-lobed; racemes greenish or purplish, drooping; fruits purplish black, glandular-bristly. **Ribes lacustre** (Pers.) Poir.; bristly black currant; Fig. 208. Moist thickets; rare.

4b.   Stems unarmed; fruit red . . . . (5)

5a.   Fruit and pedicel glandular-bristly; calyx whitish to roseate; stems up to 1 m long; leaves 5–7-lobed, cordate at the base. **Ribes glandulosum** Grauer; skunk currant; Fig. 209. Moist scrubby woodland; occasional.

5b.   Fruit smooth; pedicels with capitate glands; calyx smoke-colored to purplish; stems up to 50 cm long; bark exfoliating; leaves 3- or sometimes 5-lobed, paler below. **Ribes triste** Pall.; red currant; Fig. 210. Moist thickets and clearings; frequent.

6a.   Bracts finely glandular-ciliate; leaves deeply 3-lobed; glands usually intermixed with pubescence on the lower surface; stems up to 1 m long; nodal spines usually present; internodes often armed; fruit reddish purple. **Ribes oxyacanthoides** L.; northern gooseberry; Fig. 211. Thickets and clearings; occasional.

6b.   Bracts villous-ciliate, nonglandular; leaves nonglandular; nodal spines (when present) few and weak; internodes often unarmed. **Ribes hirtellum** Michx. Thickets and clearings; occasional.

42. ROSACEAE   rose family

1a.   Fruit consisting of 5 dry follicles splitting down one side; shrubs with simple toothed leaves . . . . . **Spiraea**
1b.   Fruit indehiscent; herbs or shrubs (or small trees) with simple or compound leaves . . . . (2)

2a.   Fruit consisting of dry achenes . . . . (3)
2b.   Fruit consisting of several-seeded pomes or 1-seeded drupes (both may be berry-like) . . . . (7)

Fig. 208. *Ribes lacustre*, 1/2 ×.

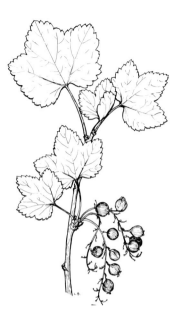

Fig. 210. *Ribes triste*, 2/5 ×.

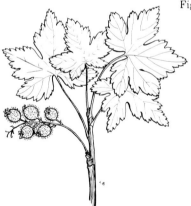

Fig. 209. *Ribes glandulosum*, 2/5 ×.

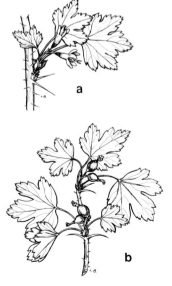

Fig. 211. *Ribes oxyacanthoides*, a, 2/5 ×; b, 2/3 ×.

3a. Achenes seed-like within the persistent calyx tube; leaves pinnately compound .... (4)
3b. Achenes superficial on the dry or fleshy receptacle .... (5)

4a. Herbs with small yellow flowers in spike-like racemes .... **Agrimonia**
4b. Shrubs with large pink or roseate flowers ...... **Rosa**

5a. Receptacle fleshy, much enlarged; leaves with 3 leaflets .... **Fragaria**
5b. Receptacle dry, little enlarged in fruit .... (6)

6a. Styles persistent and elongated in fruit, feathery or jointed; leaves pinnate; smaller leaflets alternating with the larger ones .... **Geum**
6b. Styles not elongating in fruit, mostly deciduous; leaves digitate or pinnate (if pinnate mostly without alternating smaller leaflets) .... **Potentilla**

7a. Fruit a 1-seeded drupe or a collection of drupelets; ovary superior .... (8)
7b. Fruit a large or small several-seeded pome; ovary inferior or appearing so .... (9)

8a. Fruit a collection of drupelets on a dry or spongy receptacle; leaves simple or compound .... **Rubus**
8b. Fruit consisting of solitary drupes with a bony stone; leaves simple .... **Prunus**

9a. Branches armed with long stout thorns ... **Crataegus**
9b. Branches unarmed .... (10)

10a. Leaves simple; flowers racemose ...... **Amelanchier**
10b. Leaves pinnate; flowers cymose .... **Sorbus**

Agrimonia   agrimony

Stems up to 80 cm long, sometimes branched; leaves pinnate; large-toothed leaflets alternating with very small ones; flowers small, yellow, in an elongated spiciform raceme; fruit deeply furrowed below the equatorial ring of hooked bristles. **Agrimonia striata** Michx.; agrimony. Clearings, open woods and scrub prairie; occasional.

Amelanchier juneberry, serviceberry

Shrubs or small trees up to 4 m high; leaves simple, ovate or oblong, rounded at the base, serrate towards the tip; flowers racemose; petals white, 6–9 mm long; fruit dark bluish purple, edible. ***Amelanchier alnifolia*** Nutt.; saskatoon; Fig. 212. Prairies, thickets, and borders of woods; frequent.

Crataegus hawthorn

Shrubs or small trees up to 3 m high; branches with long thorns; leaves ovate, doubly serrate; flowers corymbose; petals white; fruit scarlet. ***Crataegus chrysocarpa*** Ashe; hawthorn. Thickets and clearings; occasional.

Fragaria strawberry

Low-growing herbs, with running stems producing new plants; leaves trifoliate; leaflets broadly ovate, coarsely toothed; flowers corymbose on a scape; scape up to 30 cm long; petals white; fruit almost round; achenes in pits on the surface; calyx lobes appressed around the base. ***Fragaria virginiana*** Dcne. ssp. ***glauca*** (Wats.) Staudt; strawberry; Fig. 213. Clearings and borders of woods; frequent.

Geum avens

1a. Styles feathery, not jointed; stems up to 40 cm long; leaves mostly basal, pinnate; lobed pinnules wedge-shaped at the base; flowers usually 3, nodding; sepals purplish pink; petals pink, yellowish, or flesh-colored. ***Geum triflorum*** Pursh; three-flowered avens; Fig. 214. Scrub prairie; frequent.

1b. Styles not feathery, jointed; leaves basal and cauline .... (2)

2a. Flowers nodding, purple or flesh-colored; sepals erect or spreading; stems erect, up to 60 cm high, little branched; basal leaves lyrate-pinnate; stem leaves trifoliate. ***Geum rivale*** L.; purple avens. Open moist woodland; occasional.

2b. Flowers ascending, yellow .... (3)

3a. Upper stem leaves trifoliate; basal leaves lyrately pinnate; terminal lobe longer than the others; stems up to 120 cm long; fruiting head about 20 mm wide;

receptacles long-hirsute; styles glandless. **Geum aleppicum** Jacq.; yellow avens; Fig. 215. Clearings and scrub prairie; occasional.

3b.   Upper stem leaves not quite trifoliate; basal leaves similar to *G. aleppicum* but the terminal leaflet large and often 3-lobed; fruiting head obovoid, about 15 mm wide; receptacles glabrous or short-hispid; styles minutely glandular at the base. **Geum macrophyllum** Willd. var. **perincisum** (Rydb.) Raup; yellow avens; Fig. 216. Clearings; occasional.

Potentilla   cinquefoil, five-finger

1a.   Leaves digitate .... (2)
1b.   Leaves pinnate .... (4)

2a.   Leaflets 3 .... (3)
2b.   Leaflets 5–7, narrowly oblanceolate, serrate, white-tomentose beneath; stems up to 60 cm long, with an open cyme of yellow flowers; basal leaves long-petioled. **Potentilla gracilis** Dougl. var. **pulcherrima** (Lehm.) Fern. Clearings and scrub prairie; occasional.

3a.   Flowers white; achenes densely hairy; stems up to 20 cm long, woody at the base; leaves petiolate, dark green and shiny above; leaflets 3, narrow, wedge-shaped, with 3 teeth at the apex. **Potentilla tridentata** Ait.; three-toothed cinquefoil; Fig. 217. Open jack pine woodland; localized.

3b.   Flowers yellow; achenes glabrous; stems up to 60 cm long, branched or unbranched; leaflets obovate to elliptical, serrate; inflorescence a leafy cyme. **Potentilla norvegica** L.; rough cinquefoil; Fig. 218. Clearings, trails, and disturbed situations; frequent.

4a.   Petals and inner side of sepals dark purple; stems ascending from a decumbent woody base; leaflets 5–(7), serrate, paler below. **Potentilla palustris** (L.) Scop.; marsh cinquefoil; Fig. 219. Wet places; occasional.

4b.   Petals yellow or whitish .... (5)

5a.   Plants up to 90 cm high, glandular-villous; leaflets 7–11 on the stalked basal leaves, fewer above, ovate, serrate; cymes strict; petals white or cream-colored.

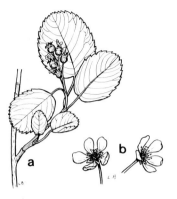

Fig. 212. *Amelanchier alnifolia*,
a, 2/5×; *b*, 4/5×.

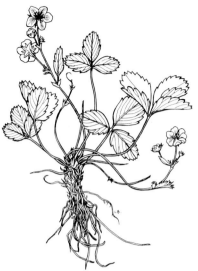

Fig. 213. *Fragaria virginiana* ssp. *glauca*, 2/5×.

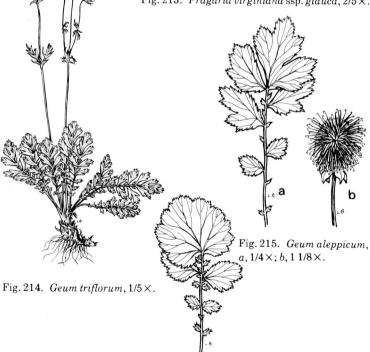

Fig. 214. *Geum triflorum*, 1/5×.

Fig. 215. *Geum aleppicum*,
a, 1/4×; *b*, 1 1/8×.

Fig. 216. *Geum macrophyllum* var. *perincisum*, 1/4×.

Potentilla arguta Pursh; white cinquefoil; Fig. 220.
Scrub prairie and disturbed situations; frequent.

5b. Plants not glandular-villous; petals yellow . . . . . . (6)

6a. Bushy shrubs up to 150 cm high, bark shreddy;
leaflets 5–7, linear-oblong, pointed at both ends,
leathery. **Potentilla fruticosa** L.; shrubby cinquefoil;
Fig. 221. Scrub prairie, clearings, and moist
woodland; common.

6b. Herbs . . . . (7)

7a. Flowers solitary on naked peduncles; low, tufted plant
spreading by runners; leaflets 7–25, green above, silky
below, often interspersed with smaller leaflets.
Potentilla anserina L.; silverweed; Fig. 222. Low wet
places; occasional.

7b. Cymes few-flowered . . . . (8)

8a. Stems up to 50 cm long; leaves mostly basal; leaflets
7–11, deeply crenate-serrate, white-tomentose below,
green to silky to grayish above; inflorescence open,
strict. **Potentilla hippiana** Lehm.; woolly cinquefoil.
Scrub prairie; occasional.

8b. Stems up to 40 cm long; leaves both basal and cauline,
with the basal having 5–7 leaflets and the cauline 3–5
leaflets; leaflets narrowly pectinate-partite, white-
tomentose below; inflorescence few-flowered to open.
Potentilla pensylvanica L. var. **bipinnatifida**
(Douglas) T. & G.; Fig. 223. Scrub prairie; rare.

Prunus   **plum, cherry**

1a. Flowers numerous in elongate racemes; shrub up to
3 m high; leaves ovate or obovate, thin, finely and
sharply serrate; fruit red purple to nearly black.
Prunus virginiana L.; choke cherry; Fig. 224. Prairies,
borders of clearings, and poplar forest; common.

1b. Flowers in small umbels or corymbs . . . . (2)

2a. Stone 12–15 mm long, more or less flattened, with a
groove on the end . . . . (3)

2b. Stone 4–5 mm long, not flattened or grooved; tree up to
8 m high; leaves ovate to lanceolate, glandular,
serrate; fruit bright red. **Prunus pensylvanica** L.f.; pin
cherry; Fig. 225. Mixed woods and borders of
clearings; common.

Fig. 217. *Potentilla tridentata*, 2/5×.

Fig. 220. *Potentilla arguta*, 1/8×.

Fig. 218. *Potentilla norvegica*, 1/4×.

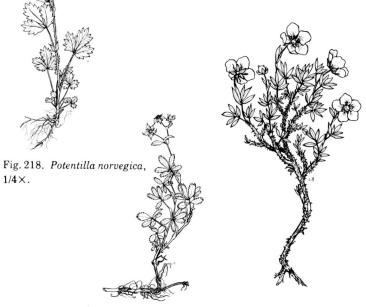

Fig. 219. *Potentilla palustris*, 1/6×.

Fig. 221. *Potentilla fruticosa*, 2/5×.

3a.   Leaves narrowly obovate with double teeth and a pointed apex; tree up to 8 m high; branches thorny; fruit red or yellow. ***Prunus americana*** Marsh.; wild plum. Mixed woods; localized dense patches and occasional single to 3 or 4 trees.

3b.   Leaves oval or obovate with rounded teeth ending in a large gland that becomes dark red in late summer; tree up to 8 m high; fruit yellow to orange. ***Prunus nigra*** Ait.; Canada plum. Mixed woods; localized.

Pyrus   apple

Small, open trees with spinescent short branchlets; leaves oblong-ovate, rounded to cordate at the base, crenate or crenate-serrate; flowers white to pink, in showy clusters. ***Pyrus malus*** L.; apple. Clearings by old habitations; rare.

Rosa   rose

1a.   Infrastipular prickles commonly present, clearly different from prickles of the internodes; stems up to 90 cm high; leaflets 5-9, elliptic-ovate, coarsely serrate; flowers 1 or several in a corymb borne on second-year branches, pink; fruit subglobose, red. ***Rosa woodsii*** Lindl.; Fig. 226. Scrub prairie; localized.

1b.   Infrastipular prickles wanting or not differentiated from prickles of the internodes . . . . (2)

2a.   Prickles extending nearly or quite to the summit of the flowering stem; stems about 1 m long; leaflets 3-7, ovate or elliptical, simply serrate; flowers 1 or in a small corymb; fruit pyriform. ***Rosa acicularis*** Lindl.; prickly rose; Fig. 227. Understory in open woods and in clearings; common.

2b.   Prickles, if any, confined to the base of the stems or only scattered above, not extending far into the flowering portion; stems up to 1.2 m long; leaflets 5-7, elliptical to oblong-obovate, serrate; flowers 1 or in a small corymb; fruit subglobose, 1.0-1.5 cm in diameter. ***Rosa blanda*** Ait.; Fig. 228. Clearings; occasional.

Rubus   raspberry, bramble

1a.   Low plants; leaves 1-3, simple, suborbicular, more or less 5-lobed; flowers single, white; fruit reddish, becoming golden yellow when ripe. ***Rubus***

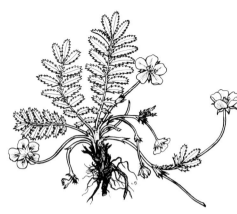

Fig. 222. *Potentilla anserina*, 2/5×.

Fig. 223. *Potentilla pensylvanica*, 1/5×.

Fig. 224. *Prunus virginiana*, a, 1/4×; b, 1/4×.

Fig. 225. *Prunus pensylvanica*, a, 1/4×; b, 1/3×.

175

chamaemorus L. cloudberry, baked-apple berry; Fig. 229. Sphagnum bogs; rare.

1b.   Leaves 3–5-foliate . . . . (2)

2a.   Stems woody, up to 2 m long, armed with bristles; leaves on sterile stems with 5 ovate, serrate leaflets that are dark green above and white-woolly below; leaves on the floricanes with 3 leaflets; flowers white, racemose; fruit red. **Rubus strigosus** Michx. (*R. idaeus* L. var. *strigosus* (Michx.) Maxim.); raspberry; Fig. 230. Open woods and clearings; common.

2b.   Stems herbaceous, unarmed . . . . (3)

3a.   Flowers usually solitary, roseate; fruit red; plants low, tufted, without runners; leaves with 3 leaflets broadly ovate and cuneate to the base. **Rubus acaulis** Michx.; stemless raspberry; Fig. 231. Bogs and low wet woods; occasional.

3b.   Flowers 1–7, white; fruit reddish purple; stems with slender runners; leaves trifoliate; leaflets ovate or rhombic, sharply toothed. **Rubus pubescens** Raf.; dewberry. Rich moist woodland; frequent.

Sorbus　　　　　　　　　　　　　　　　**mountain-ash**

Small trees up to 6 m high; leaflets 11–13, elliptical-lanceolate; flowers in dense terminal corymbs; fruit globose, red. **Sorbus decora** (Sarg.) C.K. Schneid.; mountain-ash. Open woodland; rare.

Spiraea　　　　　　　　　　　　　　　　**meadowsweet**

Shrubs up to 1 m high; leaves narrowly oblanceolate, pointed at both ends, sharply toothed; flowers white, in a terminal thyrse. **Spiraea alba** Du Roi; meadowsweet. Scrub prairie, clearings, and thickets, often low-lying; frequent.

43. LEGUMINOSAE　　pea family

1a.   Shrubs . . . . (2)
1b.   Herbs . . . . (3)

2a.   Low shrubs; flowers purple . . . . **Amorpha**
2b.   Tall shrubs; flowers yellow . . . . **Caragana**

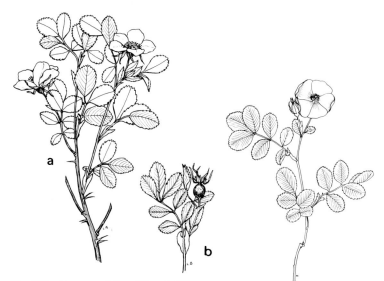

Fig. 226. *Rosa woodsii*, *a*, 2/5×; *b*, 2/5×.

Fig. 228. *Rosa blanda*, 1/4×.

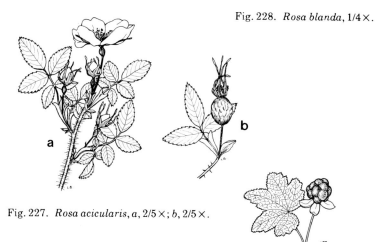

Fig. 227. *Rosa acicularis*, *a*, 2/5×; *b*, 2/5×.

Fig. 229. *Rubus chamaemorus*, 2/5×.

3a. Terminal leaflet of the pinnate leaves modified into tendrils .... (4)
3b. Terminal leaflet not modified; plants not climbing .... (5)

4a. Styles flattened, bearded down the inner face; wings essentially free .... **Lathyrus**
4b. Styles filiform, with a tuft of hairs at the summit; wings coherent with the keel .... **Vicia**

5a. Leaves palmately 3(–5)-foliate; stamens 10 ...... (6)
5b. Leaves pinnate .... (10)

6a. Leaflets toothed, at least minutely so .... (7)
6b. Leaflets entire .... (9)

7a. Pods curved or spirally coiled .... **Medicago**
7b. Pods straight or nearly so .... (8)

8a. Flowers in a loose or dense head .... **Trifolium**
8b. Flowers in slender spike-like racemes ..... **Melilotus**

9a. Stems twining .... **Amphicarpa**
9b. Stems not twining .... **Psoralea**

10a. Flowers small, in dense terminal spikes; stamens 5 .. .. **Petalostemon**
10b. Stamens 10 (9 united in a tube) .... (11)

11a. Pods jointed .... **Hedysarum**
11b. Pods continuous .... (12)

12a. Flowering stems mostly leafy; keel tips blunt .... **Astragalus**
12b. Flowering stems leafless; keel tips tapered to a sharp point .... **Oxytropis**

Amorpha   false indigo

Low branchy shrubs up to 30 cm high; leaflets 13–31, glandular-punctate below; petals purple; pod small, glandular-punctate. **Amorpha nana** Nutt.; false indigo. Steep dry bank; rare.

Amphicarpa   hog-peanut

Stems twisting and twining over other vegetation, with a ring of reflexed stiff hairs at each node; leaves trifoliolate;

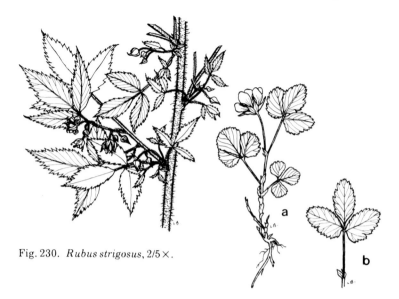

Fig. 230. *Rubus strigosus*, 2/5×.

Fig. 231. *Rubus acaulis*, a, 1/2×; b, 1/2×.

leaflets thin, ovate, pointed at the apex; raceme few-flowered, on a long peduncle; flowers whitish to pale mauve. **Amphicarpa bracteata** (L.) Fern.; hog-peanut. Moist situations below the escarpment; rare and localized.

Astragalus   **milk-vetch**

1a. Pods sessile or nearly so within the calyx . . . . (2)
1b. Pods stipitate within the calyx . . . . (6)

2a. Pods fleshy, 1.5–2 cm in diameter, indehiscent; stems spreading on the ground; leaflets 13–27, oblong to linear; inflorescence few-flowered, 4–5 cm long; flowers whitish with a purplish tip or bluish purple. **Astragalus crassicarpus** Nutt. (*A. caryocarpus* Ker); ground-plum, buffalo-bean. Scrub prairie slope; rare.
2b. Pods dry, readily dehiscent; stems ascending . . . . . (3)

3a. Calyx tube 2.5–3.5 mm long; stems up to 50 cm long, straggling; leaflets 13–23, linear to oblong; inflorescence 5–10 cm long, elongating in fruit; flowers white, tipped purplish to reddish purple; pods linear, cylindrical. **Astragalus flexuosus** Dougl. Dry sandy prairie; rare.

3b. Calyx tube 3.5–10 mm long . . . . (4)

4a. Racemes up to 20 cm long; flowers greenish white to creamy; pods glabrous; stems up to 1 m long; leaflets 13–27, elliptical to oblong. *Astragalus canadensis* L.; Fig. 232. Clearings, borders of thickets, and lakeshores; occasional.

4b. Racemes 2–4 cm long; flowers purple; pods hairy . . . . (5)

5a. Pubescence of simple hairs; stems tufted, up to 30 cm long; leaflets 11–21, lanceolate to linear-oblong; pods densely hirsute. *Astragalus agrestis* Dougl. (*A. goniatus* Nutt., *A. danicus* Retz. var. *dasyglottis* (Fisch.) Boivin). Scrub prairie, clearings, and disturbed situations; frequent.

5b. Pubescence of malpighian hairs; stems up to 40 cm long, decumbent to ascending; leaflets 9–19, elliptical to oblong; pods densely strigose. *Astragalus striatus* Nutt.; Fig. 233. Scrub prairie and disturbed situations; frequent.

6a. Corolla blue or purple . . . . (7)
6b. Corolla yellowish white; stems up to 50 cm long; leaflets 11–21, linear to linear-oblong; pods flattened, glabrous, drooping. *Astragalus tenellus* Pursh; Fig. 234. Stream banks and clearings; localized.

7a. Stems mat-forming; leaflets 11–25, oblong or oval; apex retuse or obtuse; pods reflexed, flattened, black, pubescent. *Astragalus alpinus* L.; Fig. 235. Borders of clearings; rare.

7b. Stems up to 80 cm long, ascending, densely tufted; leaflets 13–29, oblong-elliptical; pods pendant, 18–22 mm long, linear-oblong, with the upper side deeply 2-grooved. *Astragalus bisulcatus* (Hook.) Gray. Prairie clearing; rare.

Caragana   caragana

Stoloniferous shrubs up to 3 m high; leaves with an even number of leaflets; leaflets ovate, apiculate; flowers few on short shoots, yellow; pods linear; valves twisting and hanging on after the seed is shed. *Caragana arborescens* Lam. Planted as a windbreak and soil stabilizer.

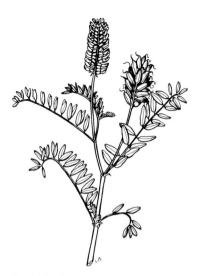

Fig. 232. *Astragalus canadensis*, 1/4×.

Fig. 234. *Astragalus tenellus*, 2/5×.

Fig. 233. *Astragalus striatus*, 2/5×.

Fig. 235. *Astragalus alpinus*, 1/2×.

Hedysarum   liquorice-root

Stems few, erect, up to 60 cm long; leaflets 9–13, lanceolate; racemes long and spike-like, terminal or axillary; flowers pinkish or violet, reflexed; segments of fruit oval, conspicuously net-veined. *Hedysarum alpinum* L. var. *americanum* Michx.; Fig. 236. Scrub prairie and roadside clearings; frequent.

Lathyrus   vetchling, wild pea

1a.   Flowers yellowish white; climbing plant with stems up to 1 m long; leaflets 6–10, oval; stipules semicordate; racemes axillary; pods up to 4 cm long. *Lathyrus ochroleucus* Hook.; pale vetchling; Fig. 237. Climbing on other vegetation in thickets and open woodland; frequent.

1b.   Flowers bluish or purple . . . . (2)

2a.   Raceme dense, shorter than the leaves, and with 15–25 purple flowers; leaflets 8–12, ovate; stems up to 100 cm long, climbing. *Lathyrus venosus* Muhl.; wild peavine. Thickets and clearings; occasional.

2b.   Raceme usually longer than the leaves, and with 2–12 blue flowers; leaflets 6–8, lanceolate to linear; stems up to 90 cm long, climbing. *Lathyrus palustris* L.; marsh vetchling. Thickets; rare.

Medicago   medick

1a.   Plants prostrate, branched; leaves trifoliolate; leaflets obovate, toothed above the middle; flowers yellow, about 3 mm long, in dense head-like racemes. *Medicago lupulina* L.; black medick. Introduced weed of roadsides and waste places; frequent.

1b.   Stems ascending, up to 80 cm long, much branched   (2)

2a.   Flowers blue to purple, 7–11 mm long; leaflets ovate to obovate, sharply toothed towards the apex; legumes coiled. *Medicago sativa* L.; alfalfa, lucerne. Escaped from cultivation; occasional.

2b.   Flowers yellow, 5–8 mm long; legumes falcate; plants otherwise much like *M. sativa*. *Medicago falcata* L. Escaped from cultivation; rare.

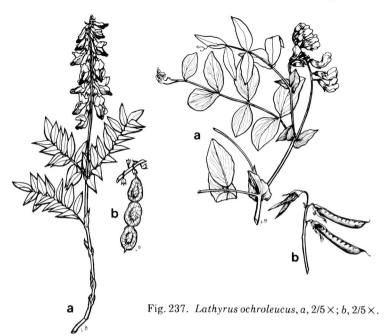

a

Fig. 237. *Lathyrus ochroleucus, a,* 2/5×; *b,* 2/5×.

b

a

Fig. 236. *Hedysarum alpinum* var. *americanum, a,* 2/5×; *b,* 4/5×.

Melilotus   sweet-clover

1a. Flowers white, 3–5 mm long; pods 3–4 mm long,
     weakly reticulate; branching stems sometimes over 2
     m; leaves trifoliate; leaflets toothed almost to the base.
     Melilotus alba Desr.; white sweet-clover. Roadsides
     and waste places; occasional.
1b. Flowers yellow, 5–6 mm long; pods 2.5–3.5 mm long,
     strongly reticulate; branching stems up to 1 m long.
     Melilotus officinalis (L.) Lam.; yellow sweet-clover.
     Roadsides and waste places; occasional.

Oxytropis   locoweed

1a. Leaflets in whorls of 3 or 4, linear-lanceolate, long
     silky pubescent; scape up to 40 cm long; inflorescence
     4–10 cm long; flowers dark pink, turning dark blue
     with age or on drying; calyx long silky pubescent.
     Oxytropis splendens Dougl.; showy locoweed; Fig.
     238. Clearings; rare.
1b. Leaflets in pairs .... (2)

183

2a.   Flowers and pods reflexed; inflorescence elongating in fruit; corolla whitish, lilac, or bluish; stems up to 40 cm long, somewhat caulescent; leaves appearing flattened; leaflets lanceolate. ***Oxytropis deflexa*** (Pall.) DC.; reflexed locoweed; Fig. 239. Scrub prairie and disturbed gravel along roadside; rare.

2b.   Flowers and pods erect or spreading; inflorescence not elongating greatly in fruit; corolla cream colored or yellow; plants definitely acaulescent; scapes up to 40 cm long; leaflets oblong-lanceolate, silky pubescent. ***Oxytropis campestris*** (L.) DC. var. ***gracilis*** (Nels.) Barneby; late yellow locoweed. Scrub prairie and borders of clearings; frequent.

Petalostemon   prairie-clover

Stems up to 50 cm long, erect or decumbent; leaflets 3–7, linear; flowers red or purple, in dense, cylindrical spikes. ***Petalostemon purpureum*** (Vent.) Rydb.; purple prairie-clover. Prairie; rare.

Psoralea   breadroot

Plants up to 60 cm high, somewhat branched, densely appressed, silvery-silky throughout; leaflets 3–5, obovate; inflorescence consisting of interrupted spikes; flowers blue, in clusters of 3 or 4. ***Psoralea argophylla*** Pursh; scurf-pea. Dry bank overlooking lake; rare.

Trifolium   clover

1a.   Flowers sessile, roseate, only the lower ones reflexed in age, occurring in a dense head 1.2–3.0 cm long; stems up to 70 cm long; leaves trifoliolate; leaflets ovate, often having a reddish inverted V on the upper surface. ***Trifolium pratense*** L.; red clover. Waste places; frequent.

1b.   Flowers pediceled in loosening heads; pedicels reflexed in age . . . . (2)

2a.   Stems widely creeping; ascending peduncles scape-like; leaflets ovate, cuneate to the base and notched at the apex, often having a whitish inverted V on the upper surface; flowers white or pinkish, in a round head-like raceme. ***Trifolium repens*** L.; white clover. Waste places; occasional.

Fig. 238. *Oxytropis spendens*, 1/5×.

Fig. 239. *Oxytropis deflexa*, a, 2/5×; b, 2/5×.

2b.   Stems up to 50 cm long, arched-ascending to erect; leaflets oval to cuneate-ovate; flowers pink, white, or roseate. ***Trifolium hybridum*** L.; alsike clover. Waste ground; occasional.

Vicia   vetch

1a.   Inflorescence consisting of 3–9 flowers; flowers in a loose raceme that is shorter than the subtending leaves; plant climbing or trailing; leaflets 8–14, elliptical or ovate, strongly veined. ***Vicia americana*** Muhl.; Fig. 240. Open woods and clearings; common.

1b.   Inflorescence consisting of 10–40 flowers; flowers in a dense raceme that equals or is longer than the subtending leaves; climbing or trailing plant; leaflets linear-oblong. ***Vicia cracca*** L.; tufted vetch. Introduced in waste places; occasional.

44. LINACEAE   flax family

Linum   flax

Erect stems up to 60 cm long; leaves linear, sharply pointed, crowded, ascending; petals 5, blue (rarely white), quickly deciduous; fruit a round capsule. **Linum lewisii** Pursh; blue flax; Fig. 241. Prairie openings, and roadside clearings; frequent.

45. OXALIDACEAE   wood-sorrel family

Oxalis   wood-sorrel

Erect or decumbent annual or perennial herbs, up to 25 cm high; stems leafy; leaves long-petioled, trifoliolate; leaflets broadly obcordate; flowers 1–9, in cymose clusters from the leaf axils, 5-parted; petals yellow, soon withering; fruit a capsule. **Oxalis stricta** L. (*O. europaea* Jordan); wooded roadside; rare.

46. GERANIACEAE   geranium family

Geranium   crane's-bill

1a.   Inflorescence loose; fruiting pedicels much longer than the calyces; beak of mature style-column 3–5 mm long; petals roseate; stems up to 50 cm long; leaves deeply dissected into narrow oblong segments. **Geranium bicknellii** Britt.; Fig. 242.   Disturbed clearings; occasional.

1b.   Inflorescence compact; fruiting pedicels about the same length as the calyces; beak of mature style-column 1–2 mm long; petals pale pink; stems up to 40 cm long; leaves deeply cut into wedge-shaped segments. **Geranium carolinianum** L.   Disturbed situations and borders of clearings; rare.

47. POLYGALACEAE   milkwort family

Polygala   milkwort

1a.   Stems up to 20 cm long; leaves scale-like below, with a few large ovate or elliptical ones above; flowers 3 or 4,

Fig. 240. *Vicia americana*, 2/5×.

Fig. 242. *Geranium bicknellii*, 1/4×.

Fig. 241. *Linum lewisii*, 1/4×.

showy, pink. **Polygala paucifolia** Willd.; fringed milkwort. Rich woods and clearings; rare.

1b. Stems up to 50 cm long, densely tufted; leaves narrowly lanceolate, finely denticulate, numerous; flowers greenish white, in a dense, terminal, spike-like raceme. **Polygala senega** L.; seneca snakeroot. Scrub prairie and roadside clearings; occasional.

48. EUPHORBIACEAE   spurge family

Euphorbia   **spurge**

1a. Stems erect up to 70 cm long; leaves alternate, linear or oblong, pointed at the apex, bluish green; flowers borne on a pair of yellowish green leaf-like bracts in an umbel-like inflorescence. **Euphorbia esula** L. s.l.; leafy spurge. Prairie-like clearing; rare.

1b. Stems prostrate; leaves opposite, broadly to narrowly oblong, dark green; flowers inconspicuous, axillary. **Euphorbia glyptosperma** Engelm. Roadside gravel; localized.

49. CALLITRICHACEAE   water-starwort family

Callitriche   **water-starwort**

Aquatic with delicate stems up to 30 cm long; leaves opposite, with the underwater ones filiform, 1-nerved and the floating ones more or less spatulate, 3-nerved; flowers minute, axillary. **Callitriche palustris** L.; water-starwort. Shallow water of ponds, marshes, and ditches, often stranded; occasional.

50. EMPETRACEAE   crowberry family

Empetrum   **crowberry**

Shrubs matted, freely branching, evergreen; leaves linear to narrowly elliptical, numerous and crowded, divergent; flowers small, purple, axillary; fruit a black berry. **Empetrum**

nigrum L. var. ***hermaphroditum*** (Lange) Sor.; black crowberry; Fig. 243. Border of open calcareous fen; localized.

51. ANACARDIACEAE   cashew family

Rhus   **sumac, poison-ivy**

Colonial shrubs up to 40 cm high, from a creeping rhizome; leaves trifoliate; leaflets ovate, entire to coarsely toothed, drooping; flowers whitish yellow, in small panicles in the leaf axils; fruit a dull white berry. ***Rhus radicans*** L. var. ***rydbergii*** (Small) Rehder; poison-ivy. Shaded ravines and lake banks; localized. May cause severe dermatitis.

52. CELASTRACEAE   stafftree family

Celastrus   **stafftree**

Twining and strangling shrubs; leaves alternate, ovate to elliptical, abruptly acuminate, serrate; inflorescence a small terminal panicle of yellowish green flowers; fruit orange, opening in age to expose bright red arils. ***Celastrus scandens*** L.; bittersweet. Moist woodland; rare.

53. ACERACEAE   maple family

Acer   **maple**

1a.   Trees up to 6 m high; leaves with 3–5 leaflets; leaflets lanceolate or ovate, toothed; flowers small, precocious, with the female ones in small greenish racemes; fruit borne in elongated clusters, often persistent well into the winter. ***Acer negundo*** L.; Manitoba maple. Mixed woods and borders of clearings; frequent.

1b.   Trees or tall shrubs; leaves simple, 3-lobed, with 3 obscure basal lobes, toothed; inflorescence an upright racemose panicle. ***Acer spicatum*** Lam.; mountain maple. Mixed woods and borders of clearings; frequent; some very dense stands on steep slopes of the escarpment.

54. BALSAMINACEAE   touch-me-not family

Impatiens   **jewelweed, touch-me-not**

1a.   Flowers orange, dotted with reddish brown or purplish spots (or unspotted in f. *immaculata* (Weath.) Fern. & Schub.); pouch sharply contracted into the reflexed spur; succulent branching annuals up to 1.5 m high; leaves ovate, bluntly toothed; fruit an explosive capsule. *Impatiens capensis* Meerb. (*I. biflora* Walt.); spotted touch-me-not; Fig. 244. Wet woods, beaver dams, and low areas by streams; occasional.

1b.   Flowers pale yellow; pouch gradually tapering to the straight spur; plant similar but lighter green. *Impatiens noli-tangere* L.; western jewelweed. Wet woods; rare.

55. RHAMNACEAE   buckthorn family

Rhamnus   **buckthorn**

Shrubs up to 2 m high; leaves ovate to elliptical, crenate, strongly veined; flowers axillary, small, greenish, single or in small umbels; fruit a black berry. *Rhamnus alnifolia* L'Her.; alder-leaved buckthorn. Wet woods and thickets; frequent. A violent laxative.

56. VITACEAE   vine family

Parthenocissus   **Virginia creeper**

Woody straggling climbers; leaves long-petioled, digitate; leaflets 5, short-petioled, ovate to broadly oblanceolate, coarsely toothed; tendrils twining, without sticky discs; flowers in panicles; fruit a few-seeded berry. *Parthenocissus inserta* (Kerner) Fritsch; Virginia creeper. Border of woodland; rare.

57. MALVACEAE   mallow family

1a.   Involucral bracts 3–9, connate; flowers terminal . . . .
. . . . *Lavatera*

Fig. 243. *Empetrum nigrum* ssp.
hermaphroditum, 1/2×.

Fig. 244. *Impatiens capensis*, 1/4×.

1b.　　Involucral bracts 3, narrow; flowers axillary . . . .
. . . . ***Malva***

Lavatera　　　　　　　　　　　　tree mallow

Erect plants up to 1.2 m high; leaves and stems densely
pubescent; lower leaves cordate-ovate; upper leaves 3-lobed,
crenate; flowers in a loose terminal raceme; petals rose pink,
deeply 2-lobed. ***Lavatera thuringiaca*** L. Roadsides; rare.

Malva　　　　　　　　　　　　　mallow

Branchy and more or less decumbent herbs; leaves deeply
cordate, broadly crenate, hirsute to stellate pubescent; flowers
axillary, in clusters of 3–5; petals white to pale mauve, about
as long as the calyx lobes. ***Malva pusilla*** Sm. (*M. rotundifolia*
of Scoggan); round-leaved mallow. Garden weed; local.

58.　HYPERICACEAE　　St. John's-wort family

Hypericum　　　　　　　　　　St. John's-wort

Stems up to 60 cm long; leaves ovate to oblong, shallowly
cordate at the base, opposite; flowers in axillary and terminal
clusters; petals pink to mauve, slightly longer than the sepals.

Hypericum virginicum L. var. **fraseri** (Spach) Fern.; marsh St. John's-wort. Floating fen; rare.

59. VIOLACEAE   violet family

Viola   **violet**

1a. Plants with elongate leafy stems; flowers axillary
     . . . . (2)
1b. Plants stemless; leaves and scapes arising from a
     rhizome or runners . . . . (4)

2a. Flowers violet; spur at least twice as long as thick;
     stems up to 30 cm long; leaves ovate with somewhat
     cordate bases. **Viola adunca** J.E. Smith; early blue
     violet; Fig. 245. Scrub prairie and clearings; common.
2b. Flowers yellow or white; spur short . . . . (3)

3a. Flowers white, often violet-tinged on the back; stems
     up to 60 cm long, with numerous stolons; leaves
     cordate, pointed at the apex. **Viola rugulosa** Greene;
     western Canada violet; Fig. 246. Mixed woods,
     clearings, and scrub prairie; frequent.
3b. Flowers yellow; stems up to 30 cm long, usually
     leafless below the middle; leaves cordate to reniform,
     mostly deltoid, crenate-serrate. **Viola pensylvanica**
     Michx. var. **leiocarpa** (Fern. & Wieg.) Fern.; smooth
     yellow violet. Moist woodland; occasional.

4a. Flowers white with purple lines; leaves orbicular-
     reniform, waxy-glossy, glabrous or essentially so.
     Viola renifolia Gray var. **brainerdii** (Greene) Fern.;
     Fig. 247. Damp woods; rare.
4b. Flowers violet or purple . . . . (5)

5a. Leaves cleft nearly to the base into three divisions;
     divisions each cleft into 2–4 lobes. **Viola pedatifida** G.
     Don; crowfoot violet. Scrub prairie; rare.
5b. Leaves merely shallowly toothed, not divided . . . . (6)

6a. Rhizome thick and fleshy . . . . (7)
6b. Rhizome slender and elongate . . . . (8)

7a. Spurred petal bearded toward the base; sepals not
     ciliate; leaves glabrous in age, cordate-ovate to
     reniform; teeth broadly flattened; petioles and

Fig. 246. *Viola rugulosa*, 1/4×.

Fig. 245. *Viola adunca*, 2/5×.

Fig. 247. *Viola renifolia* var. *brainerdii*, 2/5×.

	peduncles essentially glabrous. **Viola nephrophylla** Greene; Fig. 248. Wet meadows and swamps; rare.
7b.	Spurred petal essentially glabrous; sepals finely ciliate; leaves cordate-ovate, prominently toothed; petioles and lower surface of expanding leaves densely long-hairy. **Viola sororia** Willd. Clearings; rare.
8a.	Flowers pale violet; spur 5–8 mm long, up to two-thirds the length of the blade; petals all beardless; leaves strigose above; basal lobes converging or overlapping. **Viola selkirkii** Pursh; long-spurred violet. Wet woods; rare.
8b.	Flowers pale lilac; spur about 2 mm long; lateral petals slightly bearded; leaves glabrous; sinus open. **Viola palustris** L.; marsh violet. Wet woodland; rare.

193

60. ELAEAGNACEAE   oleaster family

1a.   Leaves alternate, silvery, undulate . . . . . . **Elaeagnus**
1b.   Leaves opposite, green above, whitened and speckled
       below, flat . . . . **Shepherdia**

Elaeagnus   oleaster

Colonial shrubs up to 1 m high; twigs scurfy; leaves oblong
or elliptical, silvery, scurfy on both sides; flowers yellowish,
foul smelling, in small axillary clusters; fruit silvery, mealy,
containing a large stony seed. **Elaeagnus commutata** Bernh.;
silverberry, wolf-willow; Fig. 249. Banks; rare.

Shepherdia   buffaloberry

Colonial shrubs up to 1 m high; twigs brown, scurfy; leaves
oval or ovate, green above and with star-shaped hairs and
rusty scales below; flowers axillary, with the male ones
clustered and the female ones single; fruit reddish or
yellowish. **Shepherdia canadensis** (L.) Nutt.; soapberry,
Canada buffaloberry; Fig. 250. Open woodland and clearings;
occasional to localized.

61. ONAGRACEAE   evening-primrose family

1a.   Fruit pear-shaped, with bristly hooked hairs, not
       splitting at maturity . . . . **Circaea**
1b.   Fruit long and narrow, lacking bristly hooked hairs,
       splitting at maturity . . . . (2)

2a.   Flowers pink, white, or purple; capsule slender; valves
       separating, recurving; seeds with a tuft of silky hairs
       at the summit . . . . **Epilobium**
2b.   Flowers yellow; capsule stout, and almost woody;
       separated valves not recurving; seeds without hairs
       . . . . **Oenothera**

Circaea   enchanter's-nightshade

Stems up to 20 cm long; leaves ovate, remotely, denticulate,
delicate; flowers small, white, in a minutely bracted raceme.

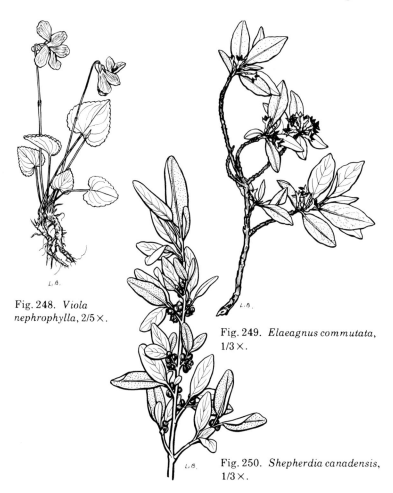

Fig. 248. *Viola nephrophylla*, 2/5×.

Fig. 249. *Elaeagnus commutata*, 1/3×.

Fig. 250. *Shepherdia canadensis*, 1/3×.

Circaea alpina L.; enchanter's-nightshade; Fig. 251. Rich moist woodland; rare.

Epilobium   willowherb

1a.   Flowers racemose; petals 1–3 cm long, pink to purple; stems up to 1.5 m long; leaves lanceolate, entire, very short-stalked. **Epilobium angustifolium** L.; fireweed, giant willowherb; Fig. 252.   Roadsides and open woodland; common.

1b.   Flowers chiefly paniculate; petals less than 1 mm long, notched at the apex .... (2)

2a.   Stems up to 1 m long, with decurrent lines running down from the leaf bases; leaves mostly opposite, lanceolate or ovate-lanceolate, more or less toothed, not revolute-margined. ***Epilobium glandulosum*** Lehm. var. ***adenocaulon*** (Haussk.) Fern.; Fig. 253. Damp places; frequent.

2b.   Stems round in cross section, without decurrent lines; leaf margins entire . . . . (3)

3a.   Stems up to 1 m long; leaves linear or linear-lanceolate, minutely hoary-pubescent, with incurved hairs; margins revolute; stolons lacking. ***Epilobium leptophyllum*** Raf.; Fig. 254. Floating fen and stony lakeshore; rare.

3b.   Stems up to 60 cm long; leaves lanceolate, glabrous or nearly so above; margins not recurved; fine stolons usually present. ***Epilobium palustre*** L.; Fig. 255. Bogs, marshes, and wet places; occasional.

Oenothera   evening-primrose

Stems up to 1 m high; leaves lanceolate to ovate-lanceolate; flowers yellow, in a leafy terminal spike; sepals reflexed; capsules cylindrical, opening at the top when mature. ***Oenothera biennis*** L.; yellow evening-primrose. Waste places; occasional.

62. HALORAGACEAE   water-milfoil family

Myriophyllum   water-milfoil

1a.   Floral bracts entire or serrate; leaves whorled, 4 or 5 per node; divisions thread-like, 4–14; turions cylindrical, tapering to a point; stems submersed, weak, whitish. ***Myriophyllum exalbescens*** Fern.; water-milfoil; Fig. 256. Quiet water 0.5–2.5 m deep; frequent.

1b.   Floral bracts pinnate or pectinate, never entire; leaves whorled, 4 or 5 per node; thread-like divisions 9–17; turions clavate; stems green. ***Myriophyllum verticillatum*** L.; water-milfoil. Quiet water 1–3 m deep; occasional.

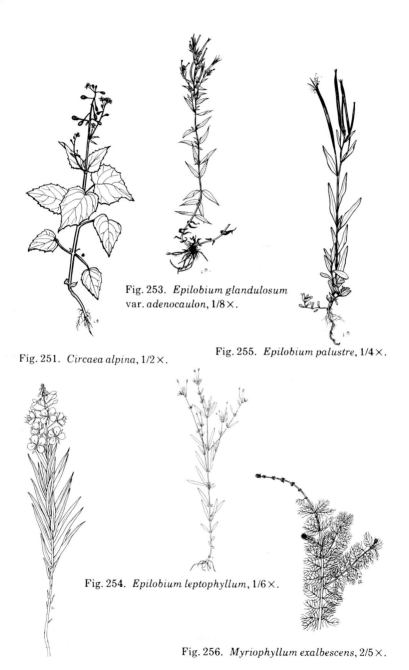

Fig. 253. *Epilobium glandulosum* var. *adenocaulon*, 1/8 ×.

Fig. 251. *Circaea alpina*, 1/2 ×.

Fig. 255. *Epilobium palustre*, 1/4 ×.

Fig. 254. *Epilobium leptophyllum*, 1/6 ×.

Fig. 256. *Myriophyllum exalbescens*, 2/5 ×.

Fig. 252. *Epilobium angustifolium*, 1/4 ×.

197

63. HIPPURIDACEAE   mare's-tail family

Hippuris   **mare's-tail**

Stems up to 50 cm long, fleshy; leaves 6–12, verticillate, linear, with those above water firm and those below water flaccid; flowers minute, occurring in the leaf axils. **Hippuris vulgaris** L.; mare's-tail; Fig. 257.   Marshes and borders of streams, ponds, and lakes; frequent.

64. ARALIACEAE   ginseng family

Aralia   **wild sarsaparilla**

Stemless herb, from a long creeping rhizome; leaves long-petioled, ternate; divisions 3–5-foliate; peduncles 20–30 cm long, bearing 1 to several umbels of greenish flowers; fruit globose, purple black, usually with 5 carpels, each carpel containing 1 seed. **Aralia nudicaulis** L.; wild sarsaparilla; Fig. 258. Moist mixed woods; common.

65. UMBELLIFERAE   parsley family

1a.	Fruit covered with hooked prickles; flowers greenish white, in compound umbels . . . . **Sanicula**
1b.	Fruit smooth or hairy; flowers white, purplish, or yellow . . . . (2)
2a.	Axils of upper leaves bearing bulblets . . . . **Cicuta**
2b.	Axils without bulblets . . . . (3)
3a.	Fruit linear or linear-oblong, 3 or more times as long as broad; flowers white . . . . **Osmorhiza**
3b.	Fruit rarely more than twice as long as broad . . . . (4)
4a.	Flowers yellow . . . . **Zizia**
4b.	Flowers white or sometimes purplish . . . . (5)
5a.	Leaves ternately compound . . . . (6)
5b.	Leaves pinnate or pinnately compound . . . . (7)
6a.	Fruit strongly flattened dorsally; lateral ribs winged; plant woolly . . . . **Heracleum**

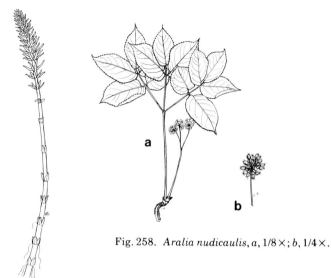

Fig. 258. *Aralia nudicaulis, a,* 1/8×; *b,* 1/4×.

Fig. 257. *Hippuris vulgaris,* 2/5×.

6b.   Fruit slightly flattened laterally, wingless; plant glabrous .... *Aegopodium*

7a.   Leaves once pinnate .... *Sium*
7b.   Leaves decompound .... (8)

8a.   Leaflets finely dissected; ultimate segments linear or filiform .... *Carum*
8b.   Leaflets merely toothed or sparingly lobed .... *Cicuta*

Aegopodium   goutweed

Erect stems up to 90 cm long; lower leaves long-petioled; leaflets oblong to ovate, sharply serrate, variegated green and white; umbels 6–15 cm wide, dense. *Aegopodium podagraria* L.; goutweed. Spread from cultivation; rare.

Carum   caraway

Stems up to 1 m long; leaves ovate, pinnately dissected; flowers white; terminal umbel usually overtopped by the lateral ones by fruiting time. *Carum carvi* L.; caraway. Roadsides; frequent.

199

Cicuta   water-hemlock

1a.   Flowers largely replaced by clusters of bulblets; stems
        up to 90 cm long; leaves dissected into segments;
        segments filiform, entire or remotely serrate. *Cicuta
        bulbifera* L.; water-hemlock; Fig. 259. Swamps;
        occasional; poisonous.
1b.   Bulblets lacking; flowers in compound umbels; stems
        up to 1.5 m long; leaflets narrowly lanceolate, sharply
        toothed. *Cicuta maculata* L. var. *angustifolia* Hook.;
        water-hemlock; Fig. 260. Swamps and borders of
        ponds; occasional; poisonous.

Heracleum   cow-parsnip

Coarse stems up to 2 m long; leaves trifoliate; leaflets up to
40 cm wide, deeply lobed, coarsely toothed; flowers white, in
flat umbels up to 30 cm wide. *Heracleum lanatum* Michx.;
cow-parsnip; Fig. 261. Wet places; occasional, localized.

Osmorhiza   sweet cicely

1a.   Styles 0.3–0.5 mm long; involucre and involucels
        wanting; stems up to 90 cm long, branched; leaflets
        thin and delicate, triangular-lanceolate, deeply cut.
        Osmorhiza depauperata Phil. (*O. obtusa* (Coult. &
        Rose) Fern.); sweet cicely. Rich low woods; rare.
1b.   Styles 2–4 mm long; involucre and involucels present;
        otherwise much like the previous species. *Osmorhiza
        longistylis* (Torrey) DC.; anise-root. Rich moist woods;
        rare.

Sanicula   snakeroot

Stems up to 1 m long; stem leaves sessile; basal leaves
long-petioled; leaflets 5–7, palmately arranged, oblanceolate,
sharply toothed; flowers greenish white, in compound umbels;
umbellets globular. *Sanicula marilandica* L.; snakeroot. Rich
moist woodland; occasional.

Sium   water-parsnip

Stems up to 1 m long or longer, hollow; leaves pinnate (or
underwater leaves 2 or 3 times pinnate); pinnae linear,
remotely toothed; flowers white, in compound umbels;
involucre of numerous lanceolate reflexed bracts. *Sium suave*

Fig. 260. *Cicuta maculata* var. *angustifolia*, *a*, 1/5×; *b*, 1/4×.

Fig. 259. *Cicuta bulbifera*, 1/6×.

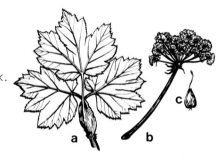

Fig. 261. *Heracleum lanatum*, *a*, 1/8×; *b*, 1/4×; *c*, 3/8×.

Walt.; water-parsnip; Fig. 262. Wet sedge meadows and borders of ponds; occasional.

Zizia   alexanders

1a.   Basal leaves simple, cordate, serrate; stem leaves ternate; leaflets ovate, with the terminal one stalked; stems up to 60 cm long; flowers bright yellow, in compound umbels. ***Zizia aptera*** (Gray) Fern.; heart-leaved alexanders. Scrub prairie and clearings; frequent.

1b.   Basal leaves ternately compound; leaflets rhomboid to lanceolate; otherwise much like *Z. aptera*. ***Zizia aurea*** (L.) Koch; golden alexanders. Scrub prairie and clearings; rare.

66. CORNACEAE   dogwood family

Cornus   **dogwood**

1a.   Inflorescence subcapitate, subtended by 4 large
       petal-like bracts; stems about 10 cm high, from a
       somewhat woody base, with 1–3 pairs of bracts and a
       verticil of 4 leaves on sterile stems and 6 leaves on
       flowering stems; leaves ovate; fruit red. **Cornus
       canadensis** L.; bunchberry; Fig. 263. Mixed woods;
       frequent.
1b.   Inflorescence cymose, without an involucre; shrubs
       . . . . (2)

2a.   Leaves ovate, alternate on leading shoots and
       subapproximate on flowering shoots; shrubs up to 2.5
       m long, with flattish tops; twigs greenish. **Cornus
       alternifolia** L. f.; green-osier, alternate-leaved
       dogwood. Border of woodland; rare.
2b.   Leaves ovate, opposite; shrubs up to 2 m high; twigs
       bright reddish colored. **Cornus stolonifera** Michx.;
       red-osier dogwood; Fig. 264. Borders of woodland and
       in low wet situations; frequent in most older stands of
       poplar forest.

67. PYROLACEAE   wintergreen family

1a.   Leaves reduced to scales, lacking chlorophyll . . . .
       . . . . **Monotropa**
1b.   Leaves green . . . . (2)

2a.   Flowers borne singly . . . . **Moneses**
2b.   Flowers racemose . . . . **Pyrola**

Moneses   **one-flowered wintergreen**

   Stems up to 15 cm long, with a single nodding waxy-white
flower; capsule erect; leaves round to ovate, crowded at the
base. **Moneses uniflora** (L.) Gray, one-flowered wintergreen;
Fig. 265. Moist woodland; occasional.

Fig. 262. *Sium suave*, 1/4×.

Fig. 264. *Cornus stolonifera*, 1/5×.

Fig. 265. *Moneses uniflora*, 2/5×.

Fig. 263. *Cornus canadensis*, 2/5×.

Monotropa   Indian-pipe

1a.   Plants waxy-white, up to 30 cm high, having a single
      nodding flower; capsule upright; plant turning black
      on drying. **Monotropa uniflora** (L.) Gray; Indian-pipe.
      Woodland; infrequent, localized in mature poplar and
      poplar–birch forests.

1b.   Plants reddish, darkening on drying; raceme few-flowered, drooping; capsules erect. ***Monotropa hypopithys*** L.; pinesap. Coniferous woodland; very rare.

Pyrola   wintergreen

1a.   Racemes 1-sided; flowers small and crowded; petals greenish; leaves broadly ovate, crenulate; stems up to 20 cm long. ***Pyrola secunda*** L.; one-sided wintergreen; Fig. 266. Moist woodland; rare.
1b.   Raceme spiral . . . . (2)

2a.   Calyx lobes rounded or obtuse, not more than 2 mm long; petals greenish white, converging; leaf blade broadly oval, usually shorter than the petiole. ***Pyrola chlorantha*** Swartz (*P. virens* Schweigg.); Fig. 267. Dry woodland; rare.
2b.   Calyx lobes lanceolate to ovate, acutish; petals spreading; leaf blade about equaling to or longer than the petiole . . . . (3)

3a.   Leaf blades elliptical to obovate, thin; petals white or creamy; stems up to 30 cm long. ***Pyrola elliptica*** Nutt.; shinleaf. Woodland; rare.
3b.   Leaves ovate, cordate at the base, leathery; petals pinkish; stems up to 30 cm long. ***Pyrola asarifolia*** Michx.; pink wintergreen; Fig. 268. Moist woodland; frequent.

68. ERICACEAE   heath family

1a.   Stems ascending . . . . (2)
1b.   Stems depressed, creeping or mat-forming . . . . (4)

2a.   Leaves densely rusty-woolly beneath . . . . ***Ledum***
2b.   Leaves not rusty-woolly beneath . . . . (3)

3a.   Leaves linear to narrowly oblong; margins strongly revolute; fruit a dry capsule . . . . ***Andromeda***
3b.   Leaves broader; margins not revolute; fruit a berry . . . . ***Vaccinium***

4a.   Corolla very deeply 4-parted; lobes reflexed . . . . ***Oxycoccus***
4b.   Corolla urn-shaped or bell-shaped . . . . (5)

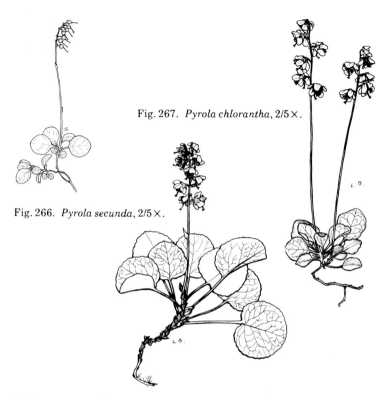

Fig. 267. *Pyrola chlorantha*, 2/5×.

Fig. 266. *Pyrola secunda*, 2/5×.

Fig. 268. *Pyrola asarifolia*, 2/5×.

5a. Flowers mostly solitary, in the leaf axils; berries bright white .... **Gaultheria**

5b. Flowers mostly in racemes or clusters, urn-shaped; fruit not white .... (6)

6a. Fruit red, blue, or blackish, tipped by the calyx teeth (ovary inferior) .... **Vaccinium**

6b. Fruit not tipped by the calyx teeth (ovary superior) .... **Arctostaphylos**

Andromeda   bog-rosemary

Stems up to 30 cm long; leaves alternate, whitened beneath, with a close minute pubescence; flowers in nodding rather dense clusters on curved branchlets; fruit a capsule. **Andromeda glaucophylla** Link; bog-rosemary. Open calcareous fen; rare.

Arctostaphylos   **bearberry**

Stems prostrate, forming large mats; leaves evergreen, spatulate; flowers pinkish white, in short few-flowered racemes; fruit red, dry, and mealy. ***Arctostaphylos uva-ursi*** (L.) Spreng.; bearberry; Fig. 269. Open woodlands, clearings, and scrub prairie; occasional.

Gaultheria   **wintergreen**

Stems prostrate, hispid; leaves short-petioled, orbicular, hispid below. ***Gaultheria hispidula*** (L.) Muhl. creeping snowberry. Moist mossy woodland; rare.

Ledum   **Labrador-tea**

Stems up to 50 cm long; leaves alternate, oblong or linear-oblong; margins inrolled, densely rusty woolly below; flowers white, in terminal umbel-like clusters; fruit a capsule. ***Ledum groenlandicum*** Oeder; Labrador-tea; Fig. 270. Bogs and spruce woodland; occasional to common.

Oxycoccus   **cranberry**

1a.   Leaves 3–5 mm long, mostly ovate; stems slender, often buried in moss; flower pedicels glabrous; berry red, 5–10 mm in diameter. ***Oxycoccus microcarpus*** Turcz.; small cranberry; Fig. 271. Bogs; apparently rather rare but perhaps overlooked.

1b.   Leaves 5–8 mm long, mostly elliptical; stems thicker, branching, and elongate; flower pedicels puberulent; berry red, 8–14 mm in diameter. ***Oxycoccus quadripetalus*** Gil.; cranberry. Open calcareous fen; rare.

Vaccinium   **blueberry, bilberry**

1a.   Stems extensively creeping; leaves evergreen, obovate; edges inrolled, dark green and shiny above, paler and glandular-dotted below; fruit red and shiny, acid. ***Vaccinium vitis-idaea*** L. var. ***minus*** Lodd.; rock cranberry or mountain cranberry; Fig. 272. Spruce woodland; occasional.

1b.   Stems upright; leaves deciduous . . . . (2)

Fig. 270. *Ledum groenlandicum*, 1/4×.

Fig. 269. *Arctostaphylos uva-ursi*, 2/5×.

Fig. 272. *Vaccinium vitis-idaea* var. *minus*, 2/3×.

Fig. 271. *Oxycoccus microcarpus*, 1×.

2a.   Leaves and twigs glabrous or nearly so; leaves oblanceolate to obovate, serrulate, sessile or nearly so; flowers axillary; berry light blue, sweet. **Vaccinium caespitosum** Michx.; dwarf bilberry; Fig. 273. Mixed woodland; rare.

2b.   Leaves and twigs pubescent; leaves oblong-lanceolate, entire; flowers in close terminal racemes; berry blue, with a bloom. **Vaccinium myrtilloides** Michx.; velvet-leaved blueberry; Fig. 274. Jack pine woodland; localized.

<h2 style="text-align:center">69. PRIMULACEAE   primrose family</h2>

1a. Plants scapose .... ***Androsace***
1b. Plants with leafy stems .... (2)

2a. Leaves opposite .... ***Lysimachia***
2b. Lower leaves small and alternate, with the upper ones
    larger and whorled .... ***Trientalis***

Androsace   pygmyflower

Leaves linear-lanceolate, toothed or entire, in a basal rosette; stems 1 to several, up to 15 cm long, each terminating in a few- to many-flowered umbel; corolla 5-lobed, constricted at the throat; capsule 5-valved. ***Androsace septentrionalis*** L.; pygmyflower; Fig. 275. Disturbed situations, roadbanks, and scrub prairie; frequent.

Lysimachia   loosestrife

1a. Flowers yellow, up to 25 mm wide, borne on long
    pedicels in the axils of the upper leaves; stems up to 80
    cm long; leaves opposite, ovate-lanceolate to ovate,
    acuminate, rounded or subcordate at the base; petioles
    ciliate-fringed. ***Lysimachia ciliata*** L. (*Steironema
    ciliata* (L.) Raf.); fringed loosestrife. Open woods and
    scrub prairie; common.
1b. Flowers small, yellow, forming short spiciform
    racemes on peduncles in the lower leaf axils; stems up
    to 50 cm long; leaves opposite, linear-lanceolate to
    lanceolate, sessile. ***Lysimachia thyrsiflora*** L.; tufted
    loosestrife; Fig. 276. Swamps, ditches, and bogs; rare.

Trientalis   starflower

Small herb up to 30 cm high; leaves 5–10, lanceolate, in a whorl; flowers few, borne on fine pedicels, white; fruit a capsule. ***Trientalis borealis*** Raf.; starflower. Rich moist woodland; rare.

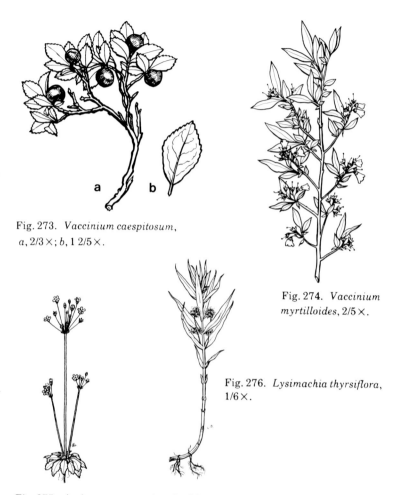

Fig. 273. *Vaccinium caespitosum*,
a, 2/3×; *b*, 1 2/5×.

Fig. 274. *Vaccinium myrtilloides*, 2/5×.

Fig. 276. *Lysimachia thyrsiflora*, 1/6×.

Fig. 275. *Androsace septentrionalis*, 2/5×.

70. OLEACEAE   olive family

Fraxinus   **ash**

Trees up to 10 m high; leaves opposite; leaflets 5–7, ovate or oblong-lanceolate; flowers inconspicuous; fruit a samara, in pendulous clusters (var. **austinii** Fern. has velvety-tomentose branchlets and pubescent petioles and leaf rachises; var.

subintegerrima (Vahl) Fern. has glabrous branchlets, petioles, and leaf rachises). **Fraxinus pennsylvanica** Marsh.; green ash; wooded slopes and along stream banks; frequent.

71. GENTIANACEAE   gentian family

1a.	Leaves basal, 3-foliate	**Menyanthes**
1b.	Leaves cauline, opposite, simple	(2)

2a.	Corolla lobes greenish or bronze, prolonged at the base into spurs	**Halenia**
2b.	Corolla lobes blue, spurless	**Gentiana**

Gentiana   **gentian**

1a. Flowers with 1 or 2 basal bracts . . . . (2)
1b. Flowers without basal bracts . . . . (3)

2a.   Leaves linear-oblong; stems up to 70 cm long; flowers 2–4, in a terminal cluster, and solitary in the upper axils; corolla 2.5–3.5 cm long, subcylindrical, with erect or slightly incurved lobes. **Gentiana rubricaulis** Schwein. (*G. linearis* Froel.); closed gentian. Marshy areas; rare.

2b.   Leaves oblong to lanceolate, firm; stems up to 30 cm long; flowers 2.5–3.0 cm long, in a dense terminal raceme-like cluster; corolla tubular, greenish blue; lobes more or less spreading at anthesis. **Gentiana affinis** Griseb. Stony lakeshore; rare.

3a.   Flowers small, 1–2 cm long; lobes acute, borne in clusters in the upper leaf axils; corolla color variable, white or yellowish to mauve or greenish or bluish; stems up to 50 cm long; upper leaves lanceolate, acute; lower leaves spatulate or ovate, blunt. **Gentiana acuta** Michx. (*G. amarella* L. var. *acuta* (Michx.) Herder); felwort; Fig. 277. Open woodland and clearings; frequent.

3b.   Flowers larger, 2–6 cm long; lobes fringed or erose-toothed . . . . (4)

4a.   Upper leaves ovate-lanceolate to ovate; stems up to 50 cm long; corolla 3.5–6.0 cm long, blue; upper half of lobes fringed. **Gentiana crinita** Froel.; fringed gentian. Wet lakeshore; rare.

4b.   Upper leaves linear-lanceolate; stems up to 50 cm long; corolla 2.3–4.0 cm long, blue; lobes with few marginal teeth. **Gentiana macounii** Holm (*G. crinita* Froel. var. *tonsa* (Lunell) Vict.; *Gentianella crinita* (Froel.) G. Don spp. *macounii* (Holm) J.M. Gillett); Fig. 278. Open calcareous fen; rare.

Halenia   spurred gentian

Stems up to 50 cm long; basal leaves spatulate or obovate; stem leaves oblong to ovate; flowers in terminal clusters and in the upper leaf axils. **Halenia deflexa** (Smith) Griseb.; spurred gentian. Moist open woodland and clearings; frequent.

Menyanthes   buck-bean

Leaves 3-foliate, alternating on a thick rhizome; leaflets large, narrowly obovate; inflorescence a raceme that terminates the naked scape; flowers white; inner side of the lobes covered with thickened hairs; capsule ellipsoid, containing numerous seeds; seeds flattened, glossy, light brown. **Menyanthes trifoliata** L.; buck-bean; Fig. 279. Shallow water and pond margins; localized.

72. APOCYNACEAE   dogbane family

Apocynum   dogbane

Stems up to 1 m long, branched, containing a milky sap; leaves opposite, paler below, ovate or oval, on a short petiole; flowers in terminal or axillary clusters, pink; lobes of the corolla recurved; fruit a pair of long narrow follicles; seeds with a tuft of hairs at the tip. **Apocynum androsaemifolium** L.; spreading dogbane; Fig. 280. Clearings and scrub prairie; occasional.

73. ASCLEPIADACEAE   milkweed family

Asclepias   milkweed

1a.   Flowers greenish white; hoods overtopping the gynostegium by about half their length; pods without tubercles; stems up to 50 cm long; leaves opposite, ovate to lanceolate, narrowing to the base; inflorescence an open umbel. ***Asclepias ovalifolia*** Dcne.; dwarf milkweed. Scrub prairie; rare.

1b. Flowers flesh-colored to pinkish purple; hoods long and lanceolate, three times longer than the stamens; pods with soft tubercles; stems to 1 m; leaves opposite, broad and oval, rounded or somewhat heart-shaped at the base; inflorescence an open umbel. ***Asclepias speciosa*** Torr.; showy milkweed. Patches in disturbed situations; rare.

74. CONVOLVULACEAE   convolvulus family

1a. Leafy twining plants with large funnel-like flowers .... ***Convolvulus***
1b. Parasitic twining plants with scale-like leaves and small flowers .... ***Cuscuta***

Convolvulus   **bindweed**

Climbing over other shrubs; leaves alternate, triangular-hastate, entire; flowers pink to white; fruit a capsule. ***Convolvulus sepium*** L.; wild morning-glory. Borders of wooded areas; occasional.

Cuscuta   **dodder**

Yellow parasitic twining plants; leaves scale-like; flowers 2–4 mm long, yellow. ***Cuscuta campestris*** Yuncker (? *C. pentagona* of auth.); dodder. Cleared areas; rare.

75. POLEMONIACEAE   phlox family

Collomia   **collomia**

Stems up to 40 cm long, more or less sticky; leaves lanceolate or linear-lanceolate, alternate; inflorescence a dense terminal leafy cluster; flowers tubular but narrow, pink or pale purple. ***Collomia linearis*** Nutt.; Fig. 281. Disturbed situations; frequent.

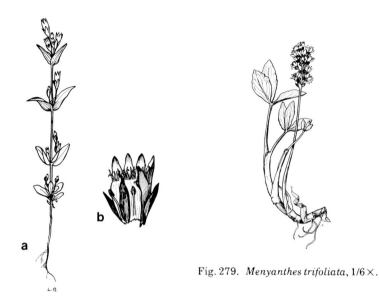

Fig. 279. *Menyanthes trifoliata*, 1/6×.

Fig. 277. *Gentiana acuta*, a, 2/5×; b, 1 1/5×.

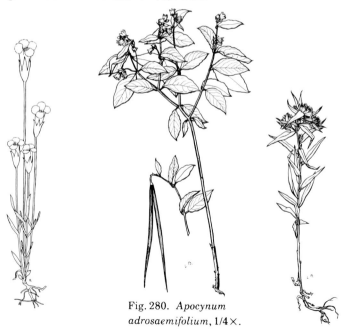

Fig. 280. *Apocynum androsaemifolium*, 1/4×.

Fig. 278. *Gentiana macounii*, 1/4×.

Fig. 281. *Collomia linearis*, 1/4×.

213

76. HYDROPHYLLACEAE   waterleaf family

Phacelia   **scorpionweed**

Stems up to 80 cm long; leaves alternate, hirsute, pinnatifid; divisions linear-oblong to triangular and often toothed; inflorescence in raceme-like cymes; calyx lobes linear, hispid; corolla bluish to whitish, rotate-campanulate. *Phacelia franklinii* (R. Br.) Gray; scorpionweed; Fig. 282. Eroding shale slope; rare.

77. BORAGINACEAE   borage family

1a.   Nutlets armed with prickles . . . . (2)
1b.   Nutlets unarmed . . . . (3)

2a.   Stem leaves linear-lanceolate . . . . *Lappula*
2b.   Stem leaves lanceolate to elliptical-oblong   *Hackelia*

3a.   Flowers yellow . . . . *Lithospermum*
3b.   Flowers blue . . . . *Mertensia*

Hackelia   **stickseed; beggar's-lice**

Stems up to 90 cm long; leaves alternate, strigose, with the lower ones petioled and the upper ones sessile; inflorescence a slender raceme; flowers small, on slender stalks, reflexed in fruit; corolla blue. *Hackelia americana* (Gray) Fern. (*Lappula deflexa* (Wahl.) Garcke var. *americana* (Gray) Greene). Clearings; rare.

Lappula   **beggarticks**

Stems up to 50 cm long, usually branched above; leaves alternate, linear-lanceolate, obtuse, sessile or somewhat petioled; flowers small, erect, in leafy-bracted racemes; corolla light blue. *Lappula echinata* Gilib.; bluebur. Roadsides and waste ground around buildings; frequent.

Lithospermum   **puccoon**

Stems up to 50 cm long, in a compact cluster from a stout rhizome; leaves alternate, linear-oblong, hoary; flowers yellow, in the axils of the upper leaves. **Lithospermum canescens** (Michx.) Lehm.; puccoon, Indian-paint. Roadsides, clearings, and scrub prairie; frequent.

Mertensia   lungwort

Stems up to 70 cm long; leaves alternate, hairy, lanceolate, with the lower ones long-petioled; inflorescence consisting of few-flowered clusters at the ends of branches; corolla purplish blue. **Mertensia paniculata** (Ait.) G. Don.; tall lungwort; Fig. 283. Open mixed woodland and clearings; frequent.

78. LABIATAE   mint family

1a.	Flowers all or mostly in one or more terminal inflorescences	(2)
1b.	Flowers axillary	(8)

2a.   Flowers in a globose head .... **Monarda**
2b.   Flowers in an elongated raceme .... (3)

3a.   Inflorescence a raceme of opposite flowers .... **Physostegia**
3b.   Inflorescence a raceme of opposite clusters .... (4)

4a.   Bracts strongly contrasted with and much shorter than the leaves .... (5)
4b.   Lower bracts grading into the upper stem leaves (6)

5a.   Leaves almost white below .... **Agastache**
5b.   Leaves green below .... **Mentha**

6a.   Upper calyx lobe at least twice as broad as any of the others .... **Dracocephalum**
6b.   Upper calyx lobe similar at least to the 2 adjacent lobes .... (7)

7a.   Flowers white .... **Nepeta**
7b.   Flowers pink or purplish .... **Stachys**

8a.   Flowers solitary in the leaf axils .... **Scutellaria**
8b.   Flowers in axillary glomerules .... (9)

9a.   Calyx strongly bilabiate .... **Dracocephalum**
9b.   Calyx weakly if at all bilabiate; lobes all similar .... (10)

10a.	Corolla weakly bilabiate .... (11)
10b.	Corolla strongly bilabiate .... (12)

11a.	Flowers sessile; stamens 2 .... **Lycopus**
11b.	Flowers pedicellate; stamens 4 .... **Mentha**

12a.	Stems creeping, rooting at the nodes ..... **Glechoma**
12b.	Stems upright .... **Galeopsis**

Agastache   giant-hyssop

Stems up to 80 cm long, branched; leaves short-pedicellate, ovate or triangular-ovate, green above, strongly whitened below, serrate; inflorescence spike-like; calyx lobes blue; corolla blue. **Agastache foeniculum** (Pursh) Ktze.; giant-hyssop; Fig. 284. Scrub prairie and clearings; frequent.

Dracocephalum   dragonhead

1a.   Inflorescence a dense, spike-like raceme; stems up to 50 cm long or longer, frequently branched; leaves petioled, lanceolate to lance-ovate, serrate; sharp teeth of bracts usually spine-tipped; corolla blue, mauve, or pink, barely longer than the calyx. **Dracocephalum parviflorum** Nutt. (*Moldavica parviflora* (Nutt.) Britton); American dragonhead; Fig. 285. Clearings; occasional.

1b.   Inflorescence in numerous axillary clusters; stems up to 50 cm long; lower leaves triangular-ovate, petioled; upper leaves ovate-lanceolate to lance-oblong, serrate; calyx glandular-dotted; upper lobe much broader than the others; corolla purplish, hardly exceeding the calyx. **Dracocephalum thymiflorum** L. Disturbed situations; rare.

Galeopsis   hemp-nettle

Stems up to 80 cm long, simple or branched, bristly-hirsute; leaves ovate, coarsely toothed, petioled; calyx lobes bristle-tipped; corolla purplish or white. **Galeopsis tetrahit** L.; hemp-nettle. Introduced weed of waste places; occasional.

Glechoma   ground-ivy

Stems up to 40 cm long, creeping; leaves round-reniform, crenate, petioled; corolla purplish blue. **Glechoma hederacea**

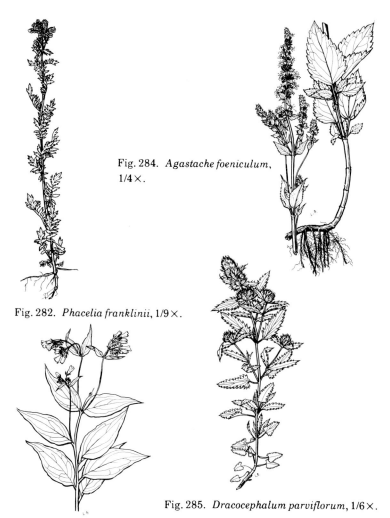

Fig. 284. *Agastache foeniculum,*
1/4×.

Fig. 282. *Phacelia franklinii,* 1/9×.

Fig. 285. *Dracocephalum parviflorum,* 1/6×.

Fig. 283. *Mertensia paniculata,* 1/6×.

L.; ground-ivy. Introduced weed of lawns and disturbed situations; rare.

Lycopus   water-horehound, bugleweed

1a.   Leaves thickish, sessile, narrowly lanceolate, with few sharp teeth on the margins; stems up to 50 cm long;

217

calyx lobes longer than the tube, acuminate. **Lycopus asper** Greene. Wet sedge meadows; rare.

1b. Leaves thin, tapered to a short or poorly defined petiole .... (2)

2a. Leaves lanceolate, short-stalked, with the lower ones deeply pinnatifid; stems up to 80 cm long; calyx teeth with long subulate tips. **Lycopus americanus** Muhl.; wet places; rare.

2b. Leaves lanceolate to lance-oblong, gradually narrowed to both ends, with widely spaced teeth; stems up to 70 cm long; calyx lobes triangular to ovate. **Lycopus uniflorus** Michx.; Fig. 286. Wet places; rare.

Mentha   mint

1a. Flowers occurring in whorls in the leaf-axils; corolla pink or mauve; stems up to 55 cm long, simple or branched; leaves ovate to lanceolate, serrate. **Mentha arvensis** L. var. **villosa** (Benth.) S.R. Stewart; field mint; Fig. 287. Wet places; common.

1b. Flowers in terminal spike-like racemes; corolla violet; stems up to 50 cm long; leaves oblong-lanceolate, sharply serrate, sessile or nearly so. **Mentha spicata** L.; spear mint. Garden escape; rare.

Monarda   wild bergamot

Stems up to 1 m long; leaves narrowly ovate to lanceolate, toothed, short-petiolate; inflorescence subtended by leaf-like bracts; corolla magenta. **Monarda fistulosa** L.; wild bergamot. Scrub prairie and clearings; frequent.

Physostegia   false dragonhead

Stems up to 1 m long; leaves sessile, oblong-lanceolate, with sharp teeth; inflorescence a terminal spike-like raceme of opposite flowers; corolla purple. **Physostegia ledinghamii** (Boivin) Cantino; false dragonhead. Moist streambanks; rare.

Prunella   selfheal

Stems simple or branched, up to 60 cm long, tufted, loosely ascending from leafy-tufted bases; leaves ovate-oblong, long petioled; inflorescence a spike-like head; 3-flowered clusters of

Fig. 286. *Lycopus uniflorus*, 1/4×.   Fig. 287. *Mentha arvensis* var. *villosa*, 1/4×.

flowers sessile in the axils of round bract-like leaves; corolla bluish, violet, or lavender. **Prunella vulgaris** L.; heal-all. Woodland trail; rare.

Scutellaria   skullcap

1a.   Flowers single or in pairs, axillary, 1.5–2.5 cm long; stems up to 60 cm long, branched or unbranched; leaves oblong to oblong-lanceolate, wavy margined; basal leaves short-petioled; upper leaves sessile; calyx with a protuberance on the upper side; corolla blue. **Scutellaria galericulata** L. var. **pubescens** Bentham.; skullcap; Fig. 288. Moist open woodland and clearings; frequent.

1b.   Flowers 5–9 mm long, in 1-sided terminal or axillary racemes; stems up to 40 cm long or longer, branched or unbranched; leaves thin, ovate, coarsely serrate or serrate-dentate; corolla blue violet. **Scutellaria lateriflora** L.; mad-dog skullcap. Moist stream bank; rare.

Stachys   hedge-nettle

Stems up to 80 cm long, branched or unbranched; leaves lanceolate to oblong-lanceolate, crenulately serrate, pubescent, sessile or nearly so; inflorescence leafy-bracted towards the base; calyx with lance-subulate teeth about equaling the tube; corolla mottled rose purple. ***Stachys palustris*** L.; woundwort; Fig. 289. Moist clearings and scrub; frequent.

79. SOLANACEAE   nightshade family

Chamaesaracha   ground-cherry

Annual with erect or ascending viscid-villous stems up to 60 cm long; leaves lance-ovate; corolla rotate, 3–5 cm wide, white with a yellow eye. ***Chamaesaracha grandiflora*** (Hook.) Fern.; large white-flowered ground-cherry. Waste ground; rare.

80. SCROPHULARIACEAE   figwort family

1a.	Leaves alternate	(2)
1b.	Leaves opposite	(5a)

2a.   Corolla yellow, spurred on the lower side at the base . . . . ***Linaria***
2b.   Corolla spurless . . . . (3)

3a.   Corolla nearly rotate, tube very short, blue or nearly white . . . . ***Veronica***
3b.   Corolla tubular or cylindrical, yellow, purple, or purple-tinged . . . . (4)

4a.   Bracts colored and showy . . . . ***Castilleja***
4b.   Bracts green . . . . ***Orthocarpus***

5a.   Anther-bearing stamens 2 . . . . ***Veronica***
5b.   Anther-bearing stamens 4 . . . . (6)

6a.   Leaves simple or slightly toothed . . . . ***Penstemon***
6b.   Leaves doubly cut-toothed . . . . ***Pedicularis***

Fig. 288. *Scutellaria galericulata* var. *pubescens*, 1/4×.

Fig. 289. *Stachys palustris*, 1/4×.

Castilleja   Indian paintbrush

1a.   Bracts broader than the leaves, scarlet or bright red; stems up to 60 cm long; leaves linear, pointed. *Castilleja miniata* Dougl.; red Indian paintbrush. Scrub prairie and clearings; frequent.

1b.   Bracts yellow, about the same width as the leaves; stems up to 50 cm long; leaves linear-lanceolate, with the upper ones sometimes shallowly lobed. *Castilleja pallida* (L.) Spreng. var. *septentrionalis* (Lindl.) Gray. Moist clearing, rare.

Linaria   toadflax

Stems up to 60 cm long; leaves alternate, linear; flowers in a terminal raceme; corolla yellow with an orange throat. *Linaria vulgaris* Miller; butter-and-eggs, yellow toadflax. Introduced weed of waste places.

221

Orthocarpus   owl's-clover

Stems up to 30 cm long, usually unbranched; leaves glandular-puberulent, linear to narrowly lanceolate, crowded, with the upper ones trifid; flowers yellow, in the axils of the upper leaves. *Orthocarpus luteus* Nutt.; owl's-clover. Scrub prairie and clearings; occasional.

Pedicularis   lousewort

Stems up to 60 cm long; leaves opposite to subopposite, oblong-lanceolate, doubly cut-toothed; inflorescence a crowded spike; corolla pale yellow. *Pedicularis lanceolata* Michx.; lousewort. Open calcareous fen; rare.

Penstemon   beardtongue

Stems up to 40 cm long; leaves opposite, linear-oblong to linear-lanceolate, slightly toothed, sessile or the basal petioled; flowers axillary, pedicellate; corolla pale purple or lilac, irregular. *Penstemon gracilis* Nutt.; beardtongue. Open pine parkland and slopes; rare.

Veronica   speedwell

1a.   Flowers solitary, in the axils of alternate upper leaves; leaves mostly sessile, spatulate to linear, with the lower ones opposite and the upper ones alternate; corolla whitish. *Veronica peregrina* L. var. *xalapensis* (HBK) St. John & Warren; neckweed, purslane, speedwell; Fig. 290. Borders of ponds and waste places; occasional.

1b.   Flowers racemose in the axils of opposite leaves . . . (2)

2a.   Leaves linear to linear-lanceolate, entire or minutely toothed; stems ascending or decumbent, up to 50 cm long; corolla blue. *Veronica scutellata* L.; marsh speedwell; Fig. 291. Swamps and sloughs; occasional.

2b.   Leaves lanceolate to narrowly ovate . . . . (3)

3a.   Leaves short-petioled; stems up to 50 cm long, decumbent; corolla blue or white. *Veronica americana* (Raf.) Schwein.; American speedwell; Fig. 292. Borders of streams; frequent.

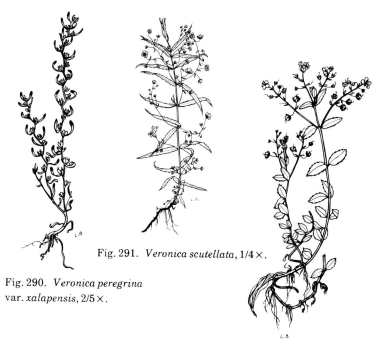

Fig. 291. *Veronica scutellata*, 1/4×.

Fig. 290. *Veronica peregrina*
var. *xalapensis*, 2/5×.

Fig. 292. *Veronica americana*, 2/5×.

3b.   Leaves sessile, at least the upper ones
      cordate-clasping; stems up to 70 cm long, decumbent
      or upright; inflorescence glandular (var. **glandulosa**
      (Farw.) Boivin) or glabrous (var. **glaberrima** (Pennell)
      Boivin). **Veronica comosa** Richter (*V. salina* of auth.);
      water speedwell. Muddy stream banks; occasional.

81. LENTIBULARIACEAE   bladderwort family

1a.   Leaves elliptical to ovate in a basal rosette ....
      .... **Pinguicula**
1b.   Leaves finely dissected, bladder-bearing **Utricularia**

Pinguicula   **butterwort**

    Leaves yellowish green, sticky, inrolled at the margin;
scape up to 10 cm long; flower single, pale purple, irregular.
Pinguicula vulgaris L.; butterwort; Fig. 293. Open calcareous
fen; rare.

223

Utricularia   **bladderwort**

1a.   Delicate plants; divisions of leaves flattened, rarely toothed, tapering to a long tip; inflorescence scapose, 3–8 mm long; flowers pale yellow. ***Utricularia minor*** L.; Fig. 294. Fens; rare.

1b.   Coarse plants; divisions of leaves round in cross section, with bristles on the margins; inflorescence scapose; flowers 1.5–2.5 cm long, yellow. ***Utricularia vulgaris*** L.; Fig. 295. Swamps and fens; occasional.

82. PLANTAGINACEAE   plantain family

Plantago   **plantain**

Leaves basal, oval to ovate, petioled, ribbed; scape up to 40 cm long; inflorescence a dense narrow spike. ***Plantago major*** L.; common plantain; Fig. 296. Introduced weed of lawns and waste places; frequent.

83. RUBIACEAE   madder family

1a.   Leaves whorled . . . . ***Galium***
1b.   Leaves opposite . . . . ***Houstonia***

Galium   **bedstraw, cleavers**

1a.   Ovary and fruit bristly or villous-hirsute . . . . (2)
1b.   Ovary and fruit smooth . . . . (4)

2a.   Principal leaves in fours, firm, 3-nerved, linear-lanceolate; stems up to 50 cm long, square, smooth; inflorescence a terminal leafy panicle. ***Galium boreale*** L. (*G. septentrionale* R. & S.); northern bedstraw; Fig. 297. Scrub prairie, clearings, and open woodland; common.

2b.   Principal leaves in sixes or eights, 1-nerved . . . . . . (3)

3a.   Stems up to 100 cm long, harsh, trailing or decumbent, annual; leaves mostly in eights, oblong-linear to oblanceolate, coarsely ciliate; cuspidate; flowers in few-flowered axillary clusters; petals white. ***Galium aparine*** L.; cleavers. Stream banks and clearings; rare.

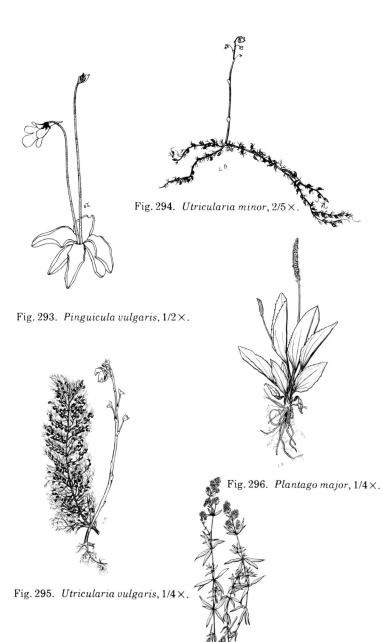

Fig. 294. *Utricularia minor*, 2/5×.

Fig. 293. *Pinguicula vulgaris*, 1/2×.

Fig. 296. *Plantago major*, 1/4×.

Fig. 295. *Utricularia vulgaris*, 1/4×.

Fig. 297. *Galium boreale*, 1/4×.

3b.   Stems up to 100 cm long, smooth, trailing or
        decumbent; leaves mostly in sixes, elliptical-
        lanceolate, finely ciliate, cuspidate; flowers in axillary
        cymes and terminal panicles; petals greenish white.
        Galium triflorum Michx.; sweet-scented bedstraw; Fig.
        298. Moist open woodland; frequent.

4a.   Corolla lobes mostly 4; stems and pedicels essentially
        smooth; stems up to 80 cm long, slender, erect or
        ascending; leaves in fours, oblanceolate or spatulate,
        soon reflexed; margins inrolled, pectinate-ciliate;
        flowers in axillary cymules. ***Galium labradoricum***
        Wieg. Marshes and fens; occasional.

4b.   Corolla lobes mostly 3; pedicels scabrous; stems up to
        40 cm long, forming dense mats; leaves in fours, linear
        or linear-oblanceolate, retrorsely scabrous-margined;
        flowers solitary, or in threes when terminal. ***Galium
        trifidum*** L.; small bedstraw; Fig. 299. Marshes and
        wet shores; frequent.

Houstonia   bluets

Stems up to 25 cm long, tufted; leaves linear to
linear-oblong, with only 2 stipules for each pair of leaves;
corolla funnel-form, pale blue; fruit an ovoid capsule.
Houstonia longifolia Gaertn. Dry scrub prairie and open jack
pine woodland; localized.

84. CAPRIFOLIACEAE   honeysuckle family

1a.   Stems slender, creeping .... ***Linnaea***
1b.   Stems stouter, erect or climbing .... (2)

2a.   Leaves entire .... (3)
2b.   Leaves toothed .... (4)

3a.   Corolla bell-shaped .... ***Symphoricarpos***
3b.   Corolla funnel-form to tubular .... ***Lonicera***

4a.   Corolla rotate; fruit a drupe with a single stone
        .... ***Virburnum***
4b.   Corolla funnel-form; fruit a slender capsule
        .... ***Diervilla***

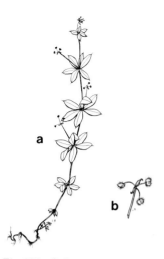

Fig. 298. *Galium triflorum, a,* 1/4×; *b,* 1 1/8×.

Fig. 299. *Galium trifidum,* 1/4×.

Diervilla
bush-honeysuckle

Shrub up to 1 m high; leaves opposite, simple, short-petioled, ovate to oval, finely toothed; flowers yellow, in small axillary and terminal clusters. ***Diervilla lonicera*** Mill.; bush-honeysuckle. Mixed woods; occasional.

Linnaea
twinflower

Leaves opposite, short-stalked, oval to orbicular, with wavy margins; flowering stems up to 10 cm long, with a pair of pendant flowers at the summit; corolla funnel-form, pink. ***Linnaea borealis*** L. var. ***americana*** (Forbes) Rehder; twinflower; Fig. 300. Moist mixed woods; occasional.

Lonicera
honeysuckle

1a.   Somewhat twining shrub; flowers yellow, in a dense terminal cluster subtended by a pair of connate leaves; leaves obovate to oval, pale below; fruit a red berry. ***Lonicera dioica*** L. var. ***glaucescens*** (Rydb.) Butt.;

227

twining honeysuckle; Fig. 301. Scrub prairie, clearings, and open woodland; frequent.

1b. Erect shrubs; flowers in pairs on axillary peduncles
. . . . (2)

2a. Flowers in pairs, subtended by 4 green to dark purple leaf-like bracts; stems up to 1 m long or longer; leaves oblong to oval; corolla yellow; berries purple to black. *Lonicera involucrata* (Richards.) Banks. Borders of woods; rare.

2b. Flowers in pairs, subtended by 2 linear bracts . . . . (3)

3a. Leaves oblong, tapering to rounded at the base, pubescent below; stems up to 1.5 m long; branchlets filled with pith; flowers yellow with a purplish tinge; berries purplish red. *Lonicera oblongifolia* (Goldie) Hook.; swamp fly honeysuckle. Wet woods; rare.

3b. Leaves ovate, more or less cordate at the base, glabrous; stems up to 3 m long; branchlets hollow at the center; flowers pink or white; berry orange or yellow. *Lonicera tatarica* L.; Tartarian honeysuckle. Escaped from or persisting after cultivation; rare.

Symphoricarpos   **snowberry**

1a. Shrub up to 60 cm high, forming large colonies; leaves thin, oval; flowers subsessile, in short axillary or terminal racemes; corolla whitish; stamens usually not exserted from the tube; berry white. *Symphoricarpos albus* (L.) Blake; snowberry; Fig. 302. Scrub prairie and clearings; infrequent.

1b. Shrub up to 1 m high or higher; leaves thicker, usually larger, oval or almost round; flowers in dense terminal or axillary spikes; corolla pink and white; styles and stamens exserted from the tube; berry white. *Symphoricarpos occidentalis* Hook.; wolfberry. Open woods, clearings, and scrub prairie; frequent.

Viburnum   **bush-cranberry**

1a. Leaves palmately 3–5 nerved from the base, mostly lobed . . . . (2)

1b. Leaves pinnately veined, unlobed . . . . (3)

2a. Flowers creamy-white, with the marginal ones sterile, much enlarged, and showy; shrubs up to 4 m high; leaves 3-lobed and more or less dentate; lobes

Fig. 300. *Linnaea borealis* var. *americana*, 1/4×.

Fig. 301. *Lonicera dioica* var. *glaucescens*, 1/4×.

Fig. 302. *Symphoricarpos albus*, 1/4×.

long-acuminate; fruit bright red. **Viburnum trilobum** Marsh. (*V. opulus* L. var. *americanum* (Mill.) Ait.); high bush-cranberry. Open woods and clearings; common.

2b. Flowers milk-white, all small and perfect; shrubs up to 1.5 m high; leaves 3–5-lobed, serrate; fruit orange-red. **Viburnum edule** (Michx.) Raf.; mooseberry, low bush-cranberry; Fig. 303. Open woods and clearings; common.

3a. Leaves ovate, finely and sharply serrate; shrubs up to 3 m high; fruit bluish black, with a bloom. **Viburnum lentago** L.; nannyberry. Clearings and open woods; occasional.

3b. Leaves oval to ovate, slightly cordate, coarsely toothed; shrubs up to 2 m high; fruit almost black. **Viburnum rafinesquianum** Schultes; downy arrowwood. Open woods and clearings; occasional.

85. VALERIANACEAE   valerian family

Valeriana   **valerian**

Stems weak, up to 60 cm long; basal leaves long-petioled, entire, spatulate; stem leaves pinnate; flowers white, in dense terminal clusters; pedicels lengthening later to form a cymose panicle. ***Valeriana septentrionalis*** Rydb. (*V. dioica* L. ssp. *sylvatica* (Sol.) Mey.); Fig. 304. Rich moist soil, scrubby clearings; occasional.

86. CAMPANULACEAE   bluebell family

Campanula   **bellflower**

1a.   Stems and leaves smooth; stems up to 50 cm long; basal leaves broadly lanceolate to deltoid or suborbicular, dentate, soon disappearing; stem leaves linear to filiform, entire; flowers 1 to several, about 2 cm long, drooping, bell-shaped, blue; fruit a capsule. ***Campanula rotundifolia*** L.; harebell, bluebell; Fig. 305. Scrub prairie, clearings, and outcrops; common.
1b.   Stems and leaves retrorsely scabrous . . . . (2)

2a.   Corolla whitish, 5–8 mm long; calyx 1.3–3.8 mm long; lobes 0.7–2.0 mm long; stems up to 60 cm long, weak; leaves lanceolate to linear-lanceolate. ***Campanula aparinoides*** Pursh; marsh bluebell. Swamps; rare.
2b.   Corolla bluish, 10–12 mm long, calyx 3.0–6.7 mm long; lobes 2.0–4.0 mm long; stems up to 60 cm long or longer; leaves narrowly linear to linear-lanceolate. ***Campanula uliginosa*** Rydb.; marsh bluebell. Swamps; rare.

87. LOBELIACEAE   lobelia family

Lobelia   **lobelia**

Stems up to 30 cm long; lower leaves spatulate; upper leaves linear; flowers racemose; corolla somewhat irregular, light blue, with a conspicuous white eye; fruit a capsule. ***Lobelia kalmii*** L.; Fig. 306. Fens and lakeshores; localized.

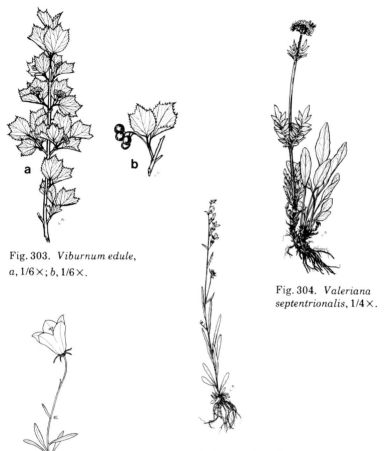

Fig. 303. *Viburnum edule,*
a, 1/6×; *b*, 1/6×.

Fig. 304. *Valeriana*
septentrionalis, 1/4×.

Fig. 306. *Lobelia kalmii*, 1/4×.

Fig. 305. *Campanula rotundifolia*, 2/5×.

88. COMPOSITAE   composite family

1a.   Heads with central tubular disc-flowers and marginal
      ligulate ray-flowers, or of disc-flowers only; juice not
      milky .... (2)

1b.   Heads with only ligulate corollas; juice usually milky
      .... (33)

231

2a. Heads rayless (or apparently so); flowers usually all tubular . . . . (3)
2b. Heads with marginal ligulate ray-flowers . . . . . . (16)

3a. Leaves prickly . . . . **Cirsium**
3b. Leaves not prickly . . . .(4)

4a. Mature fruiting heads bur-like, with hooked spines or bristles . . . . **Arctium**
4b. Mature fruiting heads not bur-like . . . . (5)

5a. Leaves with distinct leaflets, or finely dissected, or deeply cleft . . . . (6)
5b. Leaves entire, or merely toothed, or broadly lobed (9)

6a. Heads of 2 kinds, the 1-seeded fruiting ones located at the foot of slender racemes of staminate heads . . . . **Ambrosia**
6b. Heads all alike . . . . (7)

7a. Receptacle conical; odor of bruised plant like pineapple . . . . **Matricaria**
7b. Receptacle flat or merely low-convex . . . . (8)

8a. Heads corymbose . . . . **Tanacetum**
8b. Heads in spike-like, racemose, or paniculate inflorescences . . . . **Artemisia**

9a. Pappus none . . . . (10)
9b. Pappus of capillary bristles, either barbellate or plumose . . . . (11)

10a. Leaves alternate; plants mostly strong-scented . . . . **Artemisia**
10b. Leaves opposite or the upper ones alternate . . . . . .**Iva**

11a. Leaves opposite or whorled . . . . **Eupatorium**
11b. Leaves alternate . . . . (12)

12a. Phyllaries thin and papery . . . . **Antennaria**
12b. Phyllaries not thin and papery . . . . (13)

13a. Phyllaries narrowly ovate to suborbicular, in several distinct series . . . . **Liatris**
13b. Phyllaries linear to lance-attenuate, in only 1 or 2 distinct series . . . . (14)

14a. Leaves undulate to pectinate-lobed . . . . **Senecio**
14b. Leaves entire or remotely and shallowly toothed (15)

15a. Phyllaries narrow and numerous, all the same length or a few of the outer ones much shorter ..... **Erigeron**
15b. Phyllaries broader and unequal, usually imbricated, with the outer ones gradually shorter .... **Aster**

16a. Ligules yellow (sometimes purplish toward the base) .... (17)
16b. Ligules purple, blue, pink, or white .... (27)

17a. Involucre very glutinous; phyllaries with strongly recurving tips .... **Grindelia**
17b. Involucre not glutinous; phyllaries not strongly recurving at the tip .... (18)

18a. Plant aquatic; leaves dimorphic, with the submersed ones divided into capillary segments and the upper ones emersed entire to pectinate ...... **Megalodonta**
18b. Plants terrestrial; leaves not dimorphic .... (19)

19a. Pappus of (2-)4 downwardly barbed awns .... **Bidens**
19b. Pappus of bristles or scales or wanting .... (20)

20a. Pappus of capillary bristles .... (21)
20b. Pappus of chaffy scales consisting of a mere crown-like border or none .... (22)

21a. Phyllaries (at least towards the tip) thin and scarious, in several distinct series .... **Solidago**
21b. Phyllaries herbaceous, of nearly equal length ... (23)

22a. Stem leaves opposite .... **Arnica**
22b. Stem leaves alternate .... **Senecio**

23a. Leaves all or mostly opposite .... **Helianthus**
23b. Leaves all or mostly alternate .... (24)

24a. Leaves pinnately divided .... **Rudbeckia**
24b. Leaves entire or merely toothed or lobed .... (25)

25a. Ligules purplish or reddish at the base .... **Gaillardia**
25b. Ligules completely yellow .... (26)

26a. Disc flowers dark purple .... **Rudbeckia**
26b. Disc flowers yellow throughout or with the tops brown to dark purple .... **Helianthus**

27a. Leaves deeply pinnatifid to finely 2–3-divided; ligules white .... (28)

27b. Leaves entire or merely toothed or irregularly lobed .... (31)

28a. Leaves closely or coarsely dentate .... (29)
28b. Leaves finely and repeatedly dissected .... (30)

29a. Ligules about 1 mm long .... **Achillea**
29b. Ligules 3–5 cm long .... **Chrysanthemum**

30a. Ligules 3 mm long or less .... **Achillea**
30b. Ligules 4–20 mm long .... **Matricaria**

31a. Plants scapose; leaves broad, all from the rhizome .... **Petasites**
31b. Plants with leafy stems .... (32)

32a. Heads on mostly leafy branches; phyllaries in several distinct series .... **Aster**
32b. Heads on naked peduncles; phyllaries uniseriate .... **Erigeron**

33a. Plants scapose; leaves in a basal rosette .... (34)
33b. Plants with leafy stems .... (35)

34a. Achenes rough with short, hard points, slender-beaked .... **Taraxacum**
34b. Achenes smooth, beakless or short-beaked **Agoseris**

35a. Flowers yellow .... (36)
35b. Flowers whitish, pink, rose purple, or blue ...... (39)

36a. Achenes flat or flattish .... (37)
36b. Achenes columnar or nearly so, beakless .... (38)

37a. Flowers 50 or more in each head; achenes beakless .... **Sonchus**
37b. Flowers 5–20 per head; achenes with a soft filiform beak .... **Lactuca**

38a. Involucre very glandular-pilose .... **Crepis**
38b. Involucre without glands or with only a few glandular hairs .... **Hieracium**

39a. Flowers blue .... **Lactuca**
39b. Flowers whitish or pink .... **Prenanthes**

Achillea   yarrow

1a.   Leaves serrate, linear-lanceolate; stems up to 60 cm long; inflorescence corymbiform; ligules white, 4–5 mm long. ***Achillea ptarmica*** L.; sneezeweed. Escaped from cultivation; rare.

1b.   Leaves more deeply dissected . . . . (2)

2a.   Leaves linear, pinnatifid; lobes dentate; stems up to 60 cm long, simple or branching above; inflorescence corymbiform; ligules about 1 mm long, white. ***Achillea sibirica*** Ledeb. Moist thickets; occasional.

2b.   Leaves linear to linear-lanceolate, bipinnately divided; stems simple or branching above; inflorescence of numerous heads corymbiform, flat-topped; ligules 1–4 mm long, white. ***Achillea millefolium*** L. s.l. (*A. lanulosa* Nutt.); common yarrow, milfoil. Clearings, open woods, open and scrub prairies, and disturbed sites; common.

Agoseris   false dandelion

Leaves in a rosette, linear-lanceolate, occasionally toothed; scape up to 40 cm long; flowers single; ligules yellow; juice milky. ***Agoseris glauca*** (Pursh) Raf.; Fig. 307. Open and scrub prairie; frequent.

Ambrosia   ragweed

Stems up to 90 cm long, branched, from an underground rhizome; leaves opposite, pinnatifid; lobes about 5 mm wide, decurrent on the winged petiole, harsh above; male flowers in long terminal racemes, female at the base, 1-seeded, ***Ambrosia psilostachya*** DC. var. ***coronopifolia*** (T. & G.) Farwell; perennial ragweed. Waste places; rare.

Antennaria   everlasting, pussy-toes

1a.   Leaves narrow, rarely over 5 mm wide, grayish or whitish tomentose above; stems up to 30 cm long; stolons short, forming a dense carpet; tips of phyllaries often tinted sulfur yellow. ***Antennaria parvifolia*** Nutt. Prairie and clearings; occasional.

1b.   Leaves broader, green above except in youth . . . . . (2)

2a.　　Basal leaves cuneate-oblanceolate, gradually narrowed at the base, not distinctly petiolate; new rosettes not developed until fruiting time; stems up to 20 cm long; stem leaves with scarious or subulate tips. **Antennaria campestris** Rydb. (? *A. neglecta* of auth.); Fig. 308. Prairie; occasional.

2b.　　Basal leaves obovate and abruptly narrowed to a winged petiole; plants long stoloniferous; new rosettes present at flowering time. **Antennaria neodioica** Greene. Scrub prairie and clearings; frequent.

Arctium   burdock

1a.　　Flowering heads corymbose, 2–3 cm wide, densely tangled with an arachnoid tomentum; coarse plants up to 1.5 m high or higher; leaves large-petioled, roundish or ovate, mostly cordate. **Arctium tomentosum** Mill.; woolly burdock. Waste places; occasional.

1b.　　Flowering heads racemose, 1.5–2.5 cm wide, glabrous or glandular; similar to the above but usually smaller. **Arctium minus** (Hill) Bernh.; common burdock. Waste places; occasional.

Arnica   arnica

1a.　　Pappus pale yellowish brown; stems up to 70 cm long; leaves mostly in 4–5 pairs, lanceolate, remotely denticulate or entire, abundantly long villous and glandular-puberulent; flowering heads 3–5, corymbose; phyllaries broadly acute, lanate ciliate at the tip; flowers yellow. **Arnica chamissonis** Less. ssp. **foliosa** (Nutt.) Maguire; Fig. 309. Scrub prairie and clearings; occasional.

1b.　　Pappus white; leaves in 2–4 pairs . . . . (2)

2a.　　Leaves narrowly lanceolate, denticulate, with the upper ones reduced and the basal ones long-petioled; stems up to 50 cm long; heads 3–5; phyllaries somewhat glandular; flowers yellow. **Arnica lonchophylla** Greene; Fig. 310. Scrub prairie and clearings; rare.

2b.　　Leaves broader, shallowly serrate, with the upper ones often tapered to the base and the lower ones cordate, long petioled; stems up to 50 cm long; heads 1–3, with the terminal one usually larger; phyllaries somewhat glandular. **Arnica cordifolia** Hook. Open mixed deciduous woods; rare.

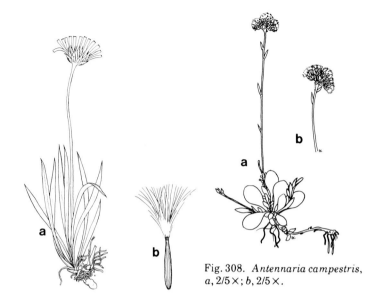

Fig. 308. *Antennaria campestris*, a, 2/5×; b, 2/5×.

Fig. 307. *Agoseris glauca*, a, 1/5×; b, 1 3/5×.

Fig. 310. *Arnica lonchophylla*, 1/4×.

Fig. 309. *Arnica chamissonis* ssp. *foliosa*, 1/6×.

237

Artemisia   wormwood

1a.   Leaves entire to coarsely lobed .... (2)
1b.   Leaves pinnatifid to tripinnatifid .... (3)

2a.   Upper leaves linear or the lower ones often trifid,
       green, and usually glabrous; stems up to 1 m long;
       panicle elongate, leafy; heads small. **Artemisia
       dracunculus** L. (*A. glauca* Pall.). Scrub prairie and
       clearings; occasional.
2b.   Leaves lanceolate, entire or slightly dentate towards
       the apex, grayish to white-tomentose on both surfaces;
       stems up to 60 cm long, branched; inflorescence a leafy
       panicle. **Artemisia ludoviciana** Nutt.; white sage.
       Scrub prairie and clearings; frequent.

3a.   Leafy segments narrow, all or mostly less than 1 mm
       wide, usually entire .... (4)
3b.   Leaf segments broader, mostly toothed or lobed ... (5)

4a.   Plants tomentose throughout, including the involucre;
       stems up to 50 cm long; inflorescence a terminal, leafy
       raceme. **Artemisia frigida** Willd. Exposed ridges and
       heavily grazed areas in grasslands; also in horse
       pastures at warden stations; infrequent.
4b.   Plants glabrous to pubescent, at least the involucre
       greenish; stems to 80 cm high, mostly unbranched;
       stem leaves short; rosette leaves up to 10 cm long or
       longer, 2–3 times divided into linear segments;
       inflorescence a leafy panicle. **Artemisia canadensis**
       Michx.; Fig. 311. Sandy lakeshores and open
       disturbed situations; rare.

5a.   Leaves green on both surfaces, the upper ones entire
       and the middle and lower ones divided; stems 1 m long,
       usually branched; inflorescence a panicle of numerous
       spike-like small heads. **Artemisia biennis** Willd.
       Waste ground and open prairies; localized.
5b.   Leaves grayish to whitish below, less so to glabrous
       above, pinnatifid to nearly tripinnatifid; stems up to 1
       m long, branched; inflorescence an ample panicle.
       Artemisia absinthium L.; absinthe, wormwood.
       Roadsides and clearings; occasional.

Aster   aster

1a.   Annual with fibrous roots; ligules of marginal florets
       wanting or rudimentary; leaves linear-attenuate,

entire; stems up to 60 cm long, somewhat branching; heads numerous. **Aster brachyactis** Blake; rayless aster. Saline meadow; rare.

1b.   Perennials with stout bases or creeping rhizomes (2)

2a.   Ligules white or sometimes pink . . . . (3)
2b.   Ligules mauve, blue, or purplish . . . . (9)

3a.   Heads in a corymb . . . . (4)
3b.   Heads in a panicle . . . . (5)

4a.   Leaves long-linear, 5 mm wide or less, with the upper ones nearly as long as the lower ones; stems slender, up to 60 cm long; inflorescence an open panicle with 1–8 heads. **Aster junciformis** Rydb.; Fig. 312. Swamps and fens; occasional.
4b.   Leaves lanceolate and much larger; stems up to 1.5 m long; flowering heads large, in a flat-topped terminal cluster. **Aster umbellatus** Mill. var. **pubens** Gray. Open woods and clearings; occasional.

5a.   Phyllaries thickish, squarrose, spinulose-mucronate; stems up to 60 cm long, branched, uniformly pubescent; leaves linear to narrowly linear-lanceolate; heads numerous, usually on 1 side of the recurved branches. **Aster ericoides** L. (*A. pansus* (Blake) Cronquist). Prairie; rare.
5b.   Phyllaries thin, straight, not mucronate; stem pubescent in lines . . . . (6)

6a.   Stems thin, with few flowering heads; leaves entire and rarely over 5 mm wide. See *A. junciformis*.
6b.   Stems stronger, usually with more than 15 heads; leaves broader . . . . (7)

7a.   Outer phyllaries larger and longer than the inner; stems up to 1 m long; leaves entire, lanceolate to narrowly linear, less than 1 cm wide; heads many, in a narrow panicle. **Aster hesperius** A. Gray (*A. johannensis* Fern.). Moist open places; occasional.
7b.   Phyllaries imbricate, with the outer ones somewhat shorter . . . . (8)

8a.   Main stem leaves usually 10–20 cm long, linear to lanceolate, remotely serrate; stems up to 1 m long or longer; heads more or less numerous, in a leafy inflorescence. **Aster simplex** Willd. Moist scrub prairie and clearings; frequent.

8b. Main stem leaves shorter and entire; panicle narrow. See *A. hesperius*.

9a. Leaves gradually dimorphic, with the lower ones petiolate .... (10)

9b. Leaves all similar, although the upper ones sometimes smaller .... (11)

10a. Leaves ovate to lanceolate, thickish and somewhat glaucous, with the lower ones cuneate to a winged petiole and the upper ones sessile with a broadly clasping base; margins scabrous; stems up to 1 m long; inflorescence stiffly racemose-paniculate to open paniculate, with greatly reduced bracteal leaves. **Aster laevis** L. Scrub prairie, clearings, and open woods; frequent.

10b. Leaves not fleshy or glaucous, with the lower ones ovate on a long and narrowly winged petiole and the upper ones narrower and stalkless; stems up to 75 cm long; panicle loosely thyrsiform. **Aster ciliolatus** Lindl.; Fig. 313. Scrub prairie and clearings; common.

11a. Stem pubescence in lines. See *A. hesperius*.

11b. Stems up to 1 m long; stem pubescence uniformly distributed; leaves not reduced upwards, long lanceolate, with bases auriculate clasping; inflorescence an ample panicle. **Aster puniceus** L. Moist scrubby clearings; frequent.

Bidens   **beggarticks**

Stems up to 80 cm long, usually branched; leaves opposite, linear, lanceolate, toothed, clasping at the base; heads nodding; flowers yellow; achenes with 4 horns. **Bidens cernua** L.; smooth beggarticks; Fig. 314. Wet lakeshores, beaver dams, pond and stream banks; common.

Chrysanthemum   **ox-eye daisy**

Stems up to 60 cm long, with few branches; leaves lyrate-pinnatifid, with the lower and basal ones petiolate and the middle ones sessile and not narrowed at the base; heads 3–5 cm wide; ray florets white; disc florets yellow. **Chrysanthemum leucanthemum** L.; ox-eye daisy. Waste places; rare.

Fig. 311. *Artemisia canadensis*, 1/5×.

Fig. 312. *Aster junciformis*, 1/4×.

Fig. 313. *Aster ciliolatus*, 1/5×.

Fig. 314. *Bidens cernua*, 1/4×.

Cirsium   thistle

1a. Flowering stems up to 1 m long, arising from extensively creeping roots; leaves sinuate-pinnatifid, with prickly margins; flowers dioecious, in large loose corymbs; florets purple. ***Cirsium arvense*** (L.) Scop.; Canada thistle. Waste places; rare.

1b. Flowering stem arising from the centre of last year's rosette; flowers perfect . . . . (2)

2a. Heads large, 5–8 cm wide, often solitary or with smaller lateral heads; inner phyllaries ending in a twisted, scarious appendage, the outer ones spine-tipped; florets purple; stems thick, up to 30 cm long, or lacking; leaves oblanceolate, green on both sides; triangular lobes with weak spines. ***Cirsium drummondii*** T. & G. Prairies; occasional.

2b. Heads smaller, more numerous; inner phyllaries narrowed to the nontwisted apex . . . . (3)

3a. Phyllaries not prickle-tipped; leaves up to 30 cm long, woolly below when young, deeply cleft into oblong or lanceolate segments that have slender spines; stems up to 1.5 m, branched; heads few, on elongate peduncle-like branches or clustered; flowers purple or white (f. ***lactiflorum*** Fern.). ***Cirsium muticum*** Michx. Moist thickets and open woods; occasional.

3b. Phyllaries glutinous on the back, prickle-tipped; leaves white-felted beneath, deeply cut into lanceolate, spiny lobes; stems up to 1 m long, branched; heads few, terminating the branches. ***Cirsium flodmanii*** (Rydb.) Arthur. Scrub prairie and clearings; rare.

Crepis   hawk's-beard

Stems up to 50 cm long, branched; stem leaves linear, sessile, with the basal and lower ones runcinate-toothed; heads terminating the branches; florets yellow. ***Crepis tectorum*** L. Roadsides and waste places; frequent.

Erigeron   fleabane

1a. Ligules inconspicuous, scarcely exceeding the disc; involucres slenderly campanulate; stems 15–90 cm long, usually branched toward the top, leaves bristly hairy, with the upper ones linear and sessile and the lower ones spatulate and short-petioled; inflorescence an open panicle with numerous small heads. ***Erigeron canadensis*** L.; horseweed. Waste places and roadsides; frequent.

1b. Ligules mostly exceeding the disc; involucres saucer-shaped to hemispherical . . . . (2)

2a. Ligules narrow, becoming involute and usually quite inconspicuous on drying . . . . (3)

2b. Ligules mostly flat and conspicuous . . . . (4)

3a. Stem leaves linear-oblong or oblanceolate; stems up to 30 cm long, somewhat hirsute; inflorescence corymb-like, with the peduncles more or less spreading or spreading-ascending, or the head solitary. ***Erigeron elatus*** (Hook.) Greene (*E. acris* L. var. *elatus* (Hook.) Cronq.); Fig. 315. Muskeg; rare.

3b. Stem leaves narrowly linear; stems up to 60 cm long, more or less hairy; inflorescence raceme-like, with the peduncles erect or nearly so, or the head sometimes solitary. ***Erigeron lonchophyllus*** Hook. Clearing by lakeshore; rare.

4a. Stem leaves cordate clasping, lanceolate, with the basal and lower ones spatulate; stems up to 60 cm long, soft; heads in a terminal corymb; florets pinkish to rose purple. ***Erigeron philadelphicus*** L.; Fig. 316. Moist clearings, lake banks and meadows; occasional.

4b. Stem leaves not clasping; stems firm . . . . (5)

5a. Plants annual, with fibrous roots; stems up to 60 cm long; leaves hispid, with the basal and lower ones ovate and petioled and the upper ones ovate to narrowly lanceolate, sharply toothed or entire; inflorescence a many-headed corymb; ligules white to lavender. ***Erigeron annuus*** (L.) Pers. (*E. ramosus* (Walt.) BSP, *E. strigosus* Muhl.). Scrub prairie and clearings; occasional.

5b. Plants perennial . . . . (6)

6a. Stems up to 40 cm long, erect at the base; basal leaves linear-oblanceolate; upper leaves linear to linear-lanceolate, becoming progressively smaller upwards; heads on ascending branches; ligules usually white. ***Erigeron asper*** Nutt. Scrub and open prairie; occasional.

6b. Stems up to 40 cm long, somewhat decumbent at the base, from a somewhat tufted rhizome; basal leaves oblanceolate; stem leaves smaller, oblong-lanceolate, pointed; heads few, on ascending branches; ligules usually purple. ***Erigeron glabellus*** Nutt.; Fig. 317. Open woods, clearings, and prairies; frequent.

Eupatorium   thoroughwort

Stems up to 1.5 m long, purplish; leaves whorled, ovate to ovate-lanceolate, acute at the apex, coarsely toothed; inflorescence a flat-topped corymb; phyllaries and florets purplish or white (f. ***faxoni*** Fern.). ***Eupatorium maculatum*** L.; spotted Joe-Pye weed. Wet places; occasional.

Gaillardia   gaillardia

Stems up to 60 cm long, erect; leaves grayish hairy; lower leaves petioled, oblong to spatulate, sometimes lobed or pinnatifid; upper leaves sessile, lanceolate, acute at the apex, entire or slightly lobed; heads large, single at the ends of branches; florets yellow at the tip and purplish towards the base. ***Gaillardia aristata*** Pursh. Scrub and open prairie; rare.

Grindelia   gumweed

Stems up to 60 cm long, branched; leaves oblanceolate, finely serrate, glandular dotted, more or less clasping; heads terminating the branches; phyllaries recurved, very gummy; florets yellow. ***Grindelia squarrosa*** (Pursh) Dunal; gumweed. Clearings just north of the park.

Helianthus   sunflower

1a.   Plants annual; stems up to 2 m long, rough; leaves alternate, ovate or deltoid-ovate, serrate; heads up to 15 cm wide; phyllaries long-caudate; ray florets yellow; disc florets dark brown or purple. ***Helianthus annuus*** L.; common sunflower. Waste places; rare.

1b.   Perennial, with rhizomes or tuberous-thickened roots; leaves all or mostly opposite . . . . (2)

2a.   Phyllaries strongly imbricate, broadly acute to rounded at the tip; stems up to 1 m long, somewhat reddish-tinged; leaves rhomboid-ovate to rhomboid-lanceolate, shallowly toothed; heads up to 8 cm wide, on long peduncles. ***Helianthus laetiflorus*** Pers. var. ***subrhomboideus*** (Rydb.) Fern. Scrub prairie and lake banks; rare.

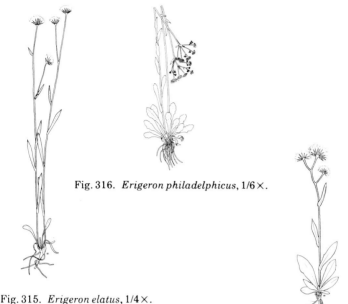

Fig. 316. *Erigeron philadelphicus*, 1/6 ×.

Fig. 315. *Erigeron elatus*, 1/4 ×.

Fig. 317. *Erigeron glabellus*, 1/4 ×.

2b.   Phyllaries narrowly acute to acuminate, somewhat loose and spreading .... (3)

3a.   Leaf blade scabrous on both sides, oblong-ovate to linear; petioles 1 cm long or shorter, ciliate; stems up to 1.5 m long or longer; heads up to 6.5 cm wide, on long peduncles. **Helianthus nuttallii** T. & G. (*H. giganteus* of auth.). Clearings and moist shores; occasional.

3b.   Leaf blades somewhat velvety below, ovate, conspicuously 3-nerved, rounded at the base; petioles 2–5 cm long; stems up to 2 m long; inflorescence corymbose, consisting of a few heads. **Helianthus tuberosus** L. var. **subcanescens** Gray. Thickets; rare.

Hieracium   hawkweed

Stems up to 1 m long; leaves ovate to lanceolate, remotely dentate, sessile; heads often subumbellate; ligules yellow.

Hieracium umbellatum L. (*H. canadense* Michx., *H. scabriusculum* Schwein.); Fig. 318. Scrub prairie, open woods, and clearings; occasional.

Iva   marsh-elder

Stems annual, up to 1.5 m long, branching; leaves broadly ovate, irregularly serrate, long-petioled, subopposite to alternate above; heads small, crowded in terminal and axillary panicles. **Iva xanthifolia** Nutt.; false ragweed. Waste places; rare.

Lactuca   lettuce

1a.   Leaves irregularly pinnatifid, coarsely toothed, with the upper ones sessile and auriculate; stems biennial, coarse, up to 2 m long or longer; inflorescence a large and dense compound panicle; heads about 5 mm wide; ligules pale or dirty blue. **Lactuca biennis** (Moench) Fern. Open moist woods; rare.

1b.   Leaves bluish, narrowly lanceolate, entire, or the lower ones remotely lobed; perennial; stems up to 1 m long; heads about 2.5 cm wide, few, in a panicle; florets bright blue. **Lactuca pulchella** (Pursh) DC.; blue lettuce; Fig. 319. Open prairie and disturbed situations; frequent.

Liatris   blazingstar

Stems stiff, up to 60 cm long; leaves linear-lanceolate, much reduced upward; heads few, in a showy terminal raceme; phyllaries erose, green, with purple tips; florets purple. **Liatris ligulistylis** (A. Nels.) K. Schum.; blazingstar. Prairie and clearings; frequent.

Matricaria   chamomile

1a.   Ray florets lacking; stems up to 40 cm long, branched; leaves numerous; segments linear; flower heads terminating the branches; crushed plants having a distinct pineapple smell. **Matricaria matricarioides** (Less.) Porter; pineappleweed. Moist situations along trails and in other disturbed sites; occasional.

1b.   Ray florets present, white; stems up to 70 cm long, branched; leaf segments thread-like; heads daisy-like,

Fig. 318. *Hieracium umbellatum*, 1/5×.

Fig. 319. *Lactuca pulchella*, 1/5×.

terminating branches. **Matricaria maritima** L. var. **agrestis** (Knaf) Wilmott. Waste places; occasional.

Megalodonta   water-marigold

Water plant with an elongated weak stem; submersed leaves sessile, ternately multifid into filiform segments, suggesting *Myriophyllum*; emergent leaves few, lanceolate-oblong; heads mostly solitary with showy yellow ligules. **Megalodonta beckii** (Torr.) Greene (*Bidens beckii* Torr.); water-marigold. Ponds; rare.

Petasites   sweet colt's-foot

1a.   Leaves unlobed oblong-cordate; margins shallowly dentate, more or less cobwebby above, strongly white-tomentose below; flowering stems precocious, bracted, up to 75 cm long in fruit; heads corymbose or corymbose-racemose; flowers whitish, with the marginal ones ligulate. **Petasites sagittatus** (Pursh) A. Gray; arrow-leaved colt's-foot; Fig. 320. Marshes and wet depressions; occasional.

247

1b.   Leaves deeply lobed, nearly glabrous above . . . . . . (2)

2a.   Leaves green and smooth above, somewhat woolly in youth, almost circular, deeply cleft almost to the base into several divisions; flowering stems much like *P. sagittatus*. **Petasites palmatus** (Ait.) A. Gray; palmate-leaved colt's-foot; Fig. 321. Moist wooded slopes and lakeshores; rare.

2b.   Leaves more or less triangular, cleft as much as half way to the midrib, green on both sides or sometimes white woolly beneath; flowering stems much like *P. sagittatus*. **Petasites vitifolius** Greene. Moist clearings and lakeshores; rare.

Prenanthes   rattlesnakeroot

1a.   Involucral bracts purplish, not hairy; stems up to 1 m long or longer; leaves deltoid, remotely dentate to deeply lobed; petioles not winged; heads drooping in a long terminal panicle; pappus cinnamon brown. **Prenanthes alba** L.; white lettuce, rattlesnakeroot. Open woodland and clearings; occasional.

1b.   Involucral bracts purplish, hirsute; stems up to 1 m long or longer; lower leaves oblanceolate to spatulate, tapering to a winged petiole; upper leaves sessile, cordate-clasping; heads in crowded clusters, spike-like; pappus straw-colored. **Prenanthes racemosa** Michx. Scrub prairie and clearings; rare.

Rudbeckia   coneflower

1a.   Leaves large, at least the middle and upper ones deeply cleft; coarse branching plants up to 2 m high; heads few, up to 10 cm wide, on long stalks; ray florets bright yellow; disc florets greenish yellow. **Rudbeckia laciniata** L.; tall coneflower. Moist ground in open woods and clearings; occasional.

1b.   Leaves entire or nearly so, lanceolate to oblanceolate; lower leaves having a winged petiole; stems up to 60 cm long, hirsute or hispid; heads single or few, on long stalks; ray florets yellow; disc florets dark brown. **Rudbeckia serotina** Nutt.; black-eyed Susan. Scrub prairie and clearings; frequent.

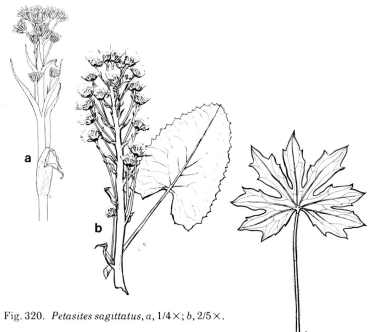

Fig. 320. *Petasites sagittatus*, *a*, 1/4×; *b*, 2/5×.

Fig. 321. *Petasites palmatus*, 2/5×.

Senecio   groundsel

1a.   Annual .... (2)
1b.   Perennial .... (3)

2a.   Stems up to 1 m long, thick, hollow; lower leaves
      lanceolate to spatulate; margins wavy; petiole winged;
      upper leaves sessile and clasping, linear-lanceolate,
      somewhat lobed or dentate; heads in 1 or more
      clusters; florets pale yellow, radiate. **Senecio
      congestus** (R. Br.) DC. (*S. palustris* (L.) Hook.);
      marsh-fleabane; Fig. 322. Wet places; rare.
2b.   Stems up to 40 cm long, branched, hollow; leaves
      oblanceolate, irregularly lobed to pinnatifid; lower
      leaves petioled; upper leaves sessile and clasping;
      flower heads with black-tipped bracts at the base,
      discoid; florets golden yellow. **Senecio vulgaris** L.;
      common groundsel. Introduced weed of waste places;
      rare.

3a.   Stems up to 1 m long; stem leaves numerous, pinnatifid; lobes narrower than the sinuses; flower heads in terminal clusters; ligules yellow. **Senecio eremophilus** Richards.; Fig. 323. Damp thickets; occasional.

3b.   Stems up to 70 cm long or longer; basal leaves round to ovate, serrate, long-petioled; stem leaves much reduced, pinnatifid; heads in a terminal cluster; ligules yellow. **Senecio aureus** L.; golden ragwort. Moist open woodland and clearings; occasional.

Solidago   **goldenrod**

1a.   Inflorescence corymbiform . . . . (2)
1b.   Inflorescence racemose, or racemose-paniculate, or in axillary clusters . . . . (3)

2a.   Leaves linear to narrowly lanceolate, sessile; stems up to 60 cm long; heads mostly sessile. **Solidago graminifolia** (L.) Salisb. var. **major** (Michx.) Fern. Borders of clearings; rare.

2b.   Leaves grayish pubescent, much broader, oval; basal leaves long-petioled; stem leaves smaller, sessile; stems up to 100 cm long or longer; heads distinctly pediceled. **Solidago rigida** L. Scrub and open prairie and clearings; frequent.

3a.   Heads spirally arranged; basal leaves much larger than the middle and upper . . . . (4)
3b.   Heads 1-sided along the upper side of the branches of the inflorescence . . . . (7)

4a.   Achenes glabrous or nearly so . . . . (5)
4b.   Achenes pubescent . . . . (6)

5a.   Leaves obovate to oblong, with the lower ones petioled and the upper ones sessile and smaller, hispid; stems up to 60 cm long, hispid; inflorescence consisting of an elongated thyrse, cylindrical or rarely narrowly paniculate. **Solidago bicolor** L. var. **concolor** T. & G. (*S. hispida* Muhl.). Open woodland, clearings, and scrub prairie; frequent.

5b.   Leaves narrowly oblanceolate, glabrous; lower leaves often scabrous margined; stems up to 40 cm long, glabrous or puberulent above; inflorescence a compact terminal panicle with erect branches. **Solidago missouriensis** Nutt. Clearings and prairie; occasional.

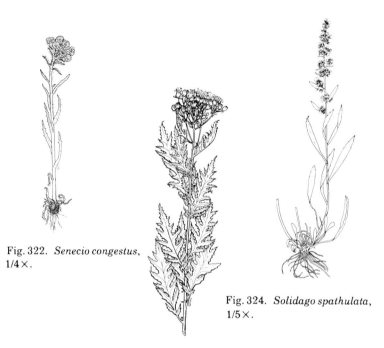

Fig. 322. *Senecio congestus*, 1/4×.

Fig. 324. *Solidago spathulata*, 1/5×.

Fig. 323. *Senecio eremophilus*, 1/4×.

6a.   Stems up to 50 cm long, decumbent at the base, commonly tufted; lower leaves spatulate, often with rounded teeth; upper leaves smaller, entire; inflorescence a narrow erect panicle. **Solidago spathulata** DC. (*S. decumbens* Greene var. *oreophila* (Rydb.) Fern.); Fig. 324. Clearings; rare.

6b.   Stems up to 40 cm long, erect from creeping rhizomes. See *Solidago missouriensis*.

7a.   Basal leaves much longer than the middle and upper ones, often forming a rosette. See *Solidago missouriensis*.

7b.   Basal leaves not much longer than the middle and upper ones, rarely forming a rosette . . . . (8)

8a.   Stems up to 1.5 m long, with the summit below the inflorescence glabrous or only sparsely pubescent; leaves lanceolate, usually sharply toothed, sessile; inflorescence large, pyramidal. **Solidago gigantea** Ait. Clearings; rare.

251

8b. Stems up to 80 cm long, with the summit below the
inflorescence densely pilose; leaves narrowly
lanceolate, finely serrate; inflorescence pyramidal.
Solidago canadensis L. Clearings and scrub prairie;
locally very abundant.

Sonchus   sow-thistle

Stems up to 1.5 m long, usually hollow and sometimes
branched; leaves runcinate-lobed, occurring mostly toward the
base; upper leaves smaller, remote; inflorescence a corymbose
panicle; flowers bright yellow. **Sonchus arvensis** L. var.
glabrescens Guenth., Grab. & Wimm. Weeds of clearings and
waste places; frequent.

Tanacetum   tansy

Stems up to 1 m long, forming dense clumps; leaves
pinnate-pinnatifid, very aromatic when bruised; heads
numerous, in a flat-topped corymb; florets yellow. **Tanacetum
vulgare** L.; tansy. Waste places; rare.

Taraxacum   dandelion

Leaves all basal, coarsely incised with triangular lobes and
a large terminal lobe, and arising from a deep fleshy taproot;
heads single, scapose; yellow florets many. **Taraxacum
officinale** Weber; dandelion. Weed of lawns and waste places;
common.

Tragopogon   goat's-beard

Stems up to 60 cm long, coarse; leaves grass-like, lanceolate
at the base, long attenuate, clasping; heads large, 3–5 cm
wide, terminating the branches; florets yellow. **Tragopogon
dubius** Scop.; yellow goat's-beard. Waste places; occasional.

Excluded species

Anthemis cotula L. — According to Scoggan (1957), the report by Lowe (1943) should be referred to *Matricaria chamomilla*. However, we could find no specimens to substantiate Scoggan's statement.

Betula glandulosa Michx. — Listed in the Riding Mountain National Park (RMNP) List of Vascular Plants. All specimens in the RMNP herbarium formerly under this name have been revised to *B. pumila* L. var. *glandulifera* Regel.

Betula occidentalis Hook. — Listed in the RMNP List of Vascular Plants. A specimen in the RMNP herbarium was revised to *B. papyrifera* Marsh.

Cardamine parviflora L. var. *arenicola* (Britt.) O.E. Schulz — Scoggan (1957) listed this species from Riding Mountain National Park on the basis of a 1948 collection by Rowe. A Rowe specimen at DAO originally determined as *C. parviflora* has been revised to *C. pensylvanica*.

Carex praegracilis Boott — Scoggan (1957) reported this species from Riding Mountain National Park on the basis of a collection by Rowe in 1948. The specimen has not been located and presumably has been revised to some other entity.

Crataegus succulenta Link — Reported in the RMNP List of Vascular Plants, as was *C. chrysocarpa*, but all the specimens we saw proved to be the latter species.

Cuscuta megalocarpa Rydb. — A sheet in the RMNP Herbarium labeled *C. megalocarpa* carried only a specimen of *Thalictrum*, the presumed host for the parasite. We found only *C. campestris* in the park.

Drosera longifolia L. — Reported by Lowe (1943) from Riding Mountain National Park. Scoggan, who could find no specimens to corroborate the record, referred it to *D. intermedia*. This is presumably what is treated as *D. anglica* in this work.

Dryopteris filix-mas (L.) Schott — Lowe (1943) recorded this species from Riding Mountain National Park, but neither Scoggan (1957) nor I have found a corroborating specimen.

Gentiana flavida Gray — According to Scoggan (1957) the report in Lowe (1943) is presumably a typographical transportation from *Halenia deflexa.*

Gentiana procera Holm — Listed in the RMNP List of Vascular Plants, but we did not find any specimens in the herbaria examined nor did we find the species during our survey. However, Scoggan (1957) reported the species from Ochre River to the northeast of the park.

Helianthus maximilianii Schrad. — Listed in the RMNP List of Vascular Plants; a specimen in the RMNP herbarium was revised to *H. nuttallii.*

Laportea canadensis (L.) Gaud. — Lowe (1943) reported this species from Riding Mountain National Park, but Scoggan (1957) did not find any specimens nor did we observe any in the park during our survey.

Pedicularis canadensis L. — Listed in the RMNP List of Vascular Plants, but we did not find any specimens in the herbaria examined nor did we observe any during our survey.

Poa trivialis L. — Lowe (1943) reported this species from Riding Mountain National Park, but neither Scoggan (1957) nor I found corroborating specimens.

Polygonum pensylvanicum L. — Scoggan (1957) reported a Gray Herbarium specimen collected by E. Scamman at Clear Lake, but the specimen could not be found in 1983 and presumably has been revised to some other taxon.

Primula incana M.E. Jones — Scoggan's (1957) report of this species from Riding Mountain is based on the listing of *P. farinosa* by Lowe (1943). No substantiating specimens have been found.

Rosa arkansana Porter — Listed in the RMNP List of Vascular Plants, but we did not find any specimens in the herbaria examined nor did we observe any during our survey; Scoggan (1957) did, however, report a specimen from "Little Saskatchewan, Man.", which he presumed was the present-day Minnedosa, to the south of the park.

Rumex crispus L. — Listed in the RMNP List of Vascular Plants, but we did not find any specimens in the herbaria examined nor did we observe any during our survey. The plants in question were probably *R. occidentalis.*

Sarracenia purpurea L. — Planted in a bog in the park a number of years ago, but presumably eliminated by a rise in the water table as a result of beaver activity.

Saxifraga tricuspidata Rottb. — Lowe (1943) sub *Chondrosea aizoon* (*S. aizoon*) reported a saxifrage from Riding Mountain National Park; Scoggan (1957) referred the report to *S. tricuspidata* on the basis of a specimen from Flin Flon, but stated that he had not seen the Riding Mountain specimen.

Selaginella rupestris (L.) Spring — Listed in the RMNP List of Vascular Plants, but no corroborating specimens have been found. The plants in question were probably *S. densa*.

Senecio tridenticulatus Rydb. — Listed in the RMNP List of Vascular Plants, but we did not see any specimens in the herbaria examined nor did we observe any during our survey.

Smilacina racemosa (L.) Desf. — Reported by Halliday (1932) and included in the RMNP List of Vascular Plants, but neither Scoggan (1957) nor I have found supporting collections.

Solidago juncea Ait. — Scoggan (1957) reported this species from Riding Mountain National Park on the basis of a 1939 collection by Heimburger. This specimen has presumably been revised to some other species. Boivin (1972) reports *S. juncea* as occurring only as far west as southeastern Manitoba.

Thaspium barbinode (Michx.) Nutt. — Listed in the RMNP List of Vascular Plants, but we did not see any specimens in the herbaria examined nor during our survey. The plants in question were probably *Zizia aurea*.

Utricularia cornuta Michx. — Lowe (1943) cited this species from Riding Mountain National Park, but Scoggan (1957) excluded it from the Flora of Manitoba because he could find neither this specimen nor any other from the province.

Valeriana officinalis L. — No specimens have been found to corroborate the Lowe (1943) report from Riding Mountain National Park.

Literature cited

Bailey, R.H. 1968. Notes on the vegetation in Riding Mountain National Park, Manitoba. National Park Forest Survey Report No. 2. Forest Management Institute, Department of Forestry and Rural Development, Canada.

Boivin, B. 1967–1981. Flora of the Prairie Provinces. Provancheria 2, 3, 4, 5, 12.

Halliday, W.E.D. 1935. Vegetation and site studies, Clear Lake, Riding Mountain National Park, Manitoba. Forest Service, Department of the Interior, Canada, Research Note No. 42.

Indian and Northern Affairs Canada. 1977. List of vascular plants, Riding Mountain National Park. Parks Canada, Indian and Northern Affairs Canada, Ottawa, Ont. 18 pp.

Lowe, C.W. 1943. List of the flowering plants, ferns, club mosses and liverworts of Manitoba. Natural History Society of Manitoba, Winnipeg, Man. 110 pp.

Porsild, A.E.; Cody, W.J. 1980. Vascular plants of continental Northwest Territories, Canada. National Museum of Natural Sciences, National Museums of Canada. 667 pp.

Rowe, S.J. 1959. Forest regions of Canada. Department of Northern Affairs and National Resources, Forestry Branch, Bull. 123. 71 pp.

Scoggan, H.J. 1957. Flora of Manitoba. Nat. Mus. Can. Bull. 140. 619 pp.

Checklist of species

Pteridophyta

1. LYCOPODIACEAE

Lycopodium annotinum
Lycopodium clavatum
   var. *monostachyon*
Lycopodium complanatum
Lycopodium dendroideum
   (*L. obscurum* pro parte)
Lycopodium lucidulum

2. SELAGINELLACEAE

Selaginella densa
Selaginella selaginoides

3. EQUISETACEAE

Equisetum arvense
Equisetum fluviatile
Equisetum hyemale ssp. *affine*
Equisetum palustre
Equisetum pratense
Equisetum scirpoides
Equisetum sylvaticum
Equisetum variegatum

4. OPHIOGLOSSACEAE

Botrychium minganense
Botrychium multifidum
Botrychium virginianum
   (ssp. *virginianum* and ssp.
europaeum)

5. PTERIDACEAE

Pteridium aquilinum var.
   latiusculum

6. ASPIDIACEAE

Athyrium filix-femina
   var. *michauxii*
Cystopteris fragilis
Dryopteris carthusiana
   (*D. spinulosa*)
Dryopteris cristata
Gymnocarpium dryopteris
   (*Dryopteris disjuncta*)
Matteuccia struthiopteris
   var. *pensylvanica*

Gymnospermae

7. PINACEAE

Abies balsamea
Juniperus communis var.
   depressa
Juniperus horizontalis
Larix laricina
Picea glauca
Picea mariana
Pinus banksiana
Thuja occidentalis

Monocotyledoneae

8. TYPHACEAE

Typha latifolia

9. SPARGANIACEAE

Sparganium angustifolium
Sparganium eurycarpum
Sparganium multipedun-
   culatum

10. POTAMOGETONACEAE

Potamogeton alpinus
  var. *tenuifolius*
Potamogeton gramineus
Potamogeton natans
Potamogeton pectinatus
Potamogeton praelongus
Potamogeton richardsonii
Potamogeton strictifolius
  var. *rutiloides*
Potamogeton vaginatus
Potamogeton zosteriformis

11. NAJADACEAE

Najas flexilis

12. SCHEUCHZERIACEAE

Scheuchzeria palustris
  var. *americana*
Triglochin maritimum
Triglochin palustre

13. ALISMACEAE

Alisma triviale
Sagittaria cuneata
Sagittaria latifolia

14. HYDROCHARITACEAE

Elodea canadensis (*Anacharis canadensis*)

15. GRAMINEAE

×*Agrohordeum macounii*
  (*Agropyron trachycaulum*
  × *Hordeum jubatum*)
Agropyron cristatum
Agropyron repens
Agropyron smithii

Agropyron trachycaulum [var. *trachycaulum*, var. *novae-angliae*, var. *glaucum*, and var. *unilaterale* (*A. subsecundum*)]
Agrostis scabra
Agrostis stolonifera (*A. alba*)
Alopecurus aequalis
Andropogon gerardi
Avena fatua
Avena sativa
Beckmannia syzigachne
Bromus ciliatus
Bromus inermis
Bromus latiglumis
Bromus porteri
Bromus pumpellianus
Calamagrostis canadensis
Calamagrostis inexpansa
Calamagrostis neglecta
Cinna latifolia
Danthonia intermedia
Danthonia spicata
Deschampsia caespitosa
Echinochloa wiegandii
Elymus canadensis
Elymus diversiglumis (*E. interruptus*)
Elymus innovatus
Elymus virginicus var. *submuticus*
Festuca hallii (*F. scabrella* pro parte)
Festuca pratensis (*F. elatior*)
Festuca rubra
Festuca saximontana
Glyceria borealis
Glyceria grandis
Glyceria striata
Helictotrichon hookeri (*Avena hookeri*)
Hierochloe odorata
Hordeum jubatum
Koeleria macrantha (*K. cristata*)
Lolium multiflorum
Lolium perenne
Milium effusum var. *cistatlanticum*

Muhlenbergia andina
Muhlenbergia cuspidata
Muhlenbergia glomerata
Muhlenbergia mexicana
Muhlenbergia racemosa
Muhlenbergia richardsonis
Oryzopsis asperifolia
Oryzopsis canadensis
Oryzopsis pungens
Phalaris arundinacea
Phleum pratense
Phragmites australis
Poa annua
Poa arida
Poa compressa
Poa nemoralis
  (incl. *P. interior*)
Poa palustris
Poa pratensis (*P. agassizensis*)
Puccinellia distans
Schizachne purpurascens
Scolochloa festucacea
Setaria viridis
Spartina gracilis
Sphenopholis intermedia
Sporobolus heterolepis
Stipa richardsonii
Stipa spartea var. *curtiseta*
Stipa viridula
Torreyochloa pallida
  var. *fernaldii*
Triticum aestivum
Triticum turgidum

16. CYPERACEAE

Carex adusta
Carex alopecoidea
Carex aquatilis
Carex assiniboinensis
Carex atherodes
Carex aurea
Carex backii
Carex bebbii
Carex brunnescens
Carex capillaris
Carex castanea
Carex chordorrhiza

Carex concinna
Carex curta (*C. canescens*)
Carex deflexa
Carex deweyana
Carex diandra
Carex disperma
Carex granularis
Carex gynocrates
Carex hookeriana
Carex houghtoniana
Carex hystricina
Carex interior
Carex lacustris
Carex lanuginosa
Carex lasiocarpa var.
  americana
Carex leptalea
Carex limosa
Carex magellanica
  (*C. paupercula*)
Carex microptera
  (*C. festivella*)
Carex obtusata
Carex peckii
Carex pedunculata
Carex pensylvanica
Carex prairea
Carex praticola
Carex pseudo-cyperus
Carex retrorsa
Carex richardsonii
Carex rosea
Carex rossii
Carex rostrata
Carex sartwellii
Carex siccata
Carex sprengelii
Carex sterilis
Carex stipata
Carex sychnocephala
Carex tenera
Carex tenuiflora
Carex torreyi
Carex trisperma
Carex vaginata
Carex viridula
Carex vulpinoidea
Carex xerantica
Eleocharis acicularis

259

Eleocharis palustris
Eleocharis pauciflora
Eleocharis smallii
Eleocharis uniglumis
Eriophorum angustifolium
Eriophorum chamissonis
Eriophorum gracile
Eriophorum spissum
Eriophorum viridi-carinatum
Scirpus caespitosus
   ssp. *austriacus*
Scirpus cyperinus
Scirpus microcarpus
   (*S. rubrotinctus*)
Scirpus validus

17. ARACEAE

Acorus calamus
Calla palustris

18. LEMNACEAE

Lemna minor
Lemna trisulca
Spirodela polyrhiza
Wolffia columbiana

19. JUNCACEAE

Juncus alpinus
Juncus balticus var. *littoralis*
Juncus bufonius
Juncus compressus
Juncus dudleyi
Juncus filiformis
Juncus nodosus
Luzula multiflora
Luzula pilosa var. *americana*
   (*L. acuminata*)

20. LILIACEAE

Allium schoenoprasum
   var. *sibiricum*

Allium stellatum
Disporum trachycarpum
Lilium philadelphicum
   (var. *philadelphicum* and
   var. *andinum*)
Maianthemum canadense
   var. *interius*
Smilacina stellata
Smilacina trifolia
Smilax herbacea
   var. *lasioneuron*
Tofieldia glutinosa
Trillium cernuum
Zygadenus elegans

21. IRIDACEAE

Sisyrinchium montanum

22. ORCHIDACEAE

Calypso bulbosa
Corallorhiza maculata
Corallorhiza striata
Corallorhiza trifida
Cypripedium calceolus
   (var. *parviflorum* and
   var. *pubescens*)
Goodyera repens
Habenaria dilatata
   (*Platanthera dilatata*)
Habenaria hyperborea
   (*Platanthera hyperborea*)
Habenaria obtusata
   (*Platanthera obtusata*)
Habenaria orbiculata
   (*Platanthera orbiculata*)
Habenaria viridis var.
   bracteata (*Coeloglossum*
   bracteatum)
Liparis loeselii
Listera cordata
Orchis rotundifolia
   (*Amerorchis rotundifolia*)
Spiranthes lacera
Spiranthes romanzoffiana

Dicotyledoneae

23. SALICACEAE

Populus balsamifera
Populus × jackii
  (*P. balsamifera × deltoides*)
Populus tremuloides
Salix amygdaloides
Salix bebbiana
Salix candida
Salix discolor
Salix fragilis
Salix gracilis (*S. petiolaris*)
Salix interior
Salix lucida
Salix lutea
Salix maccalliana
Salix myrtillifolia
Salix padophylla
  (*S. pseudomonticola*)
Salix pedicellaris
  var. *hypoglauca*
Salix pellita
Salix planifolia
Salix pyrifolia
Salix serissima

24. BETULACEAE

Alnus crispa
Alnus incana ssp. *rugosa*
Betula papyrifera
Betula pumila var.
  glandulifera (*B. glandulosa*
  var. *glandulifera*)
Corylus americana
Corylus cornuta

25. FAGACEAE

Quercus macrocarpa

26. ULMACEAE

Ulmus americana

27. CANNABACEAE

Humulus lupulus

28. URTICACEAE

Urtica dioica ssp. *gracilis*

29. SANTALACEAE

Comandra umbellata
  (*C. pallida*)
Geocaulon lividum

30. POLYGONACEAE

Polygonum achoreum
Polygonum amphibium
Polygonum aviculare
Polygonum cilinode
Polygonum convolvulus
Polygonum douglasii
Polygonum hydropiper
Polygonum lapathifolium
Polygonum persicaria
Polygonum scandens
Rumex fennicus
Rumex occidentalis
Rumex orbiculatus
Rumex maritimus
  var. *fueginus*
Rumex triangulivalvis
  (*R. mexicanus*)

31. CHENOPODIACEAE

Atriplex subspicata
Axyris amaranthoides
Chenopodium album
Chenopodium berlandieri ssp.
  zschackei
Chenopodium capitatum
Chenopodium giganto-
  spermum
Chenopodium glaucum

Chenopodium leptophyllum
Chenopodium pratericola
Chenopodium rubrum
Chenopodium strictum
   var. *glaucophyllum*
Monolepis nuttalliana

32. AMARANTHACEAE

Amaranthus graecizans
Amaranthus retroflexus

33. PORTULACACEAE

Portulaca oleracea

34. CARYOPHYLLACEAE

Cerastium arvense
Cerastium nutans
Dianthus deltoides
Gypsophila paniculata
Lychnis chalcedonica
Minuartia dawsonensis
   (*Arenaria dawsonensis*)
Moehringia lateriflora
   (*Arenaria lateriflora*)
Saponaria officinalis
Silene alba
Silene drummondii (*Lychnis*
   drummondii, L. pudica)
Silene noctiflora
Silene vulgaris (*S. cucubalus*)
Stellaria calycantha
Stellaria crassifolia
Stellaria longifolia
Stellaria longipes
Stellaria media

35. CERATOPHYLLACEAE

Ceratophyllum demersum

36. NYMPHAEACEAE

Nuphar microphyllum
Nuphar variegatum

37. RANUNCULACEAE

Actaea rubra (f. *rubra* and
   f. *neglecta*)
Anemone canadensis
Anemone cylindrica
Anemone multifida
Anemone quinquefolia
Anemone virginiana
   (*A. riparia*)
Aquilegia brevistyla
Aquilegia canadensis
Caltha palustris
Coptis trifolia
Delphinium glaucum
Pulsatilla ludoviciana
   (*Anemone patens*)
Ranunculus abortivus
Ranunculus acris
Ranunculus aquatilis
   var. *subrigidus*
Ranunculus cymbalaria
Ranunculus gmelinii
Ranunculus lapponicus
Ranunculus macounii
Ranunculus pensylvanicus
Ranunculus rhomboideus
Ranunculus sceleratus
Thalictrum dasycarpum
Thalictrum venulosum

38. FUMARIACEAE

Corydalis aurea

39. CRUCIFERAE

Arabis divaricarpa
Arabis drummondii

Arabis glabra
Arabis hirsuta
    var. pycnocarpa
Brassica campestris
Capsella bursa-pastoris
Cardamine pensylvanica
Descurainia richardsonii
Descurainia sophia
Draba nemorosa var. leiocarpa
Erucastrum gallicum
Erysimum cheiranthoides
Erysimum inconspicuum
Hesperis matronalis
Lepidium densiflorum
Rorippa islandica
Sinapis arvensis (Brassica
    kaber var. pinnatifida)
Sisymbrium altissimum
Thlaspi arvense

40. DROSERACEAE

Drosera anglica
Drosera rotundifolia

41. SAXIFRAGACEAE

Chrysosplenium alternifolium
    var. ioense
Heuchera richardsonii
Mitella nuda
Parnassia glauca
Parnassia palustris
    var. neogaea (P. multiseta)
Ribes americanum
    (R. floridum)
Ribes glandulosum
Ribes hirtellum
Ribes hudsonianum
Ribes lacustre
Ribes oxyacanthoides
Ribes triste

42. ROSACEAE

Agrimonia striata
Amelanchier alnifolia

Crataegus chrysocarpa
Fragaria virginiana
Geum aleppicum var. strictum
Geum macrophyllum
    var. perincisum
Geum rivale
Geum triflorum
Potentilla anserina
Potentilla arguta
Potentilla fruticosa
Potentilla gracilis
    var. pulcherrima
Potentilla hippiana
Potentilla norvegica
Potentilla palustris
Potentilla pensylvanica
    var. bipinnatifida
Potentilla tridentata
Prunus americana
Prunus nigra
Prunus pensylvanica
Prunus virginiana
Pyrus malus
Rosa acicularis
Rosa blanda
Rosa woodsii
Rubus acaulis
Rubus chamaemorus
Rubus pubescens
Rubus strigosus (R. idaeus)
Sorbus decora
Spiraea alba

43. LEGUMINOSAE

Amorpha nana
Amphicarpa bracteata
Astragalus agrestis
    (A. danicus var.
    dasyglottis, A. goniatus)
Astragalus alpinus
Astragalus bisulcatus
Astragalus canadensis
Astragalus crassicarpus
Astragalus flexuosus
Astragalus striatus
    (A. adsurgens
    var. robustior)

Astragalus tenellus
Caragana arborescens
Hedysarum alpinum
    var. *americanum*
Lathyrus ochroleucus
Lathyrus palustris
Lathyrus venosus
Medicago falcata
Medicago lupulina
Medicago sativa
Melilotus alba
Melilotus officinalis
Oxytropis campestris
    var. *gracilis*
Oxytropis deflexa var. *sericea*
Oxytropis splendens
Petalostemon purpureum
Psoralea argophylla
Trifolium hybridum
Trifolium pratense
Trifolium repens
Vicia americana
Vicia cracca

44. LINACEAE

Linum lewisii

45. OXALIDACEAE

Oxalis stricta

46. GERANIACEAE

Geranium bicknellii
Geranium carolinianum

47. POLYGALACEAE

Polygala paucifolia
Polygala senega

48. EUPHORBIACEAE

Euphorbia esula
Euphorbia glyptosperma

49. CALLITRICHACEAE

Callitriche palustris

50. EMPETRACEAE

Empetrum nigrum
    var. *hermaphroditum*

51. ANACARDIACEAE

Rhus radicans

52. CELASTRACEAE

Celastrus scandens

53. ACERACEAE

Acer negundo
Acer spicatum

54. BALSAMINACEAE

Impatiens capensis (f. *capensis*
    and f. *immaculata*)
Impatiens noli-tangere

55. RHAMNACEAE

Rhamnus alnifolia

56. VITACEAE

Parthenocissus inserta

57. MALVACEAE

Lavatera thuringiaca
Malva pusilla

58. HYPERICACEAE

Hypericum virginicum
    var. *fraseri*

59. VIOLACEAE

Viola adunca
Viola nephrophylla
Viola palustris
Viola pedatifida
Viola pensylvanica
    var. *leiocarpa*
Viola renifolia var. *brainerdii*
Viola rugulosa
Viola selkirkii
Viola sororia

60. ELAEAGNACEAE

Elaeagnus commutata
Shepherdia canadensis

61. ONAGRACEAE

Circaea alpina
Epilobium angustifolium
Epilobium glandulosum
    var. *adenocaulon*
Epilobium leptophyllum
Epilobium palustre
Oenothera biennis

62. HALORAGACEAE

Myriophyllum exalbescens
Myriophyllum verticillatum

63. HIPPURIDACEAE

Hippuris vulgaris

64. ARALIACEAE

Aralia nudicaulis

65. UMBELLIFERAE

Aegopodium podagraria
Carum carvi
Cicuta bulbifera
Cicuta maculata
    var. *angustifolia*
Heracleum lanatum
Osmorhiza depauperata
    (*O. obtusa*)
Osmorhiza longistylis
Sanicula marilandica
Sium suave
Zizia aptera
Zizia aurea

66. CORNACEAE

Cornus alternifolia
Cornus canadensis
Cornus stolonifera

67. PYROLACEAE

Moneses uniflora
Monotropa hypopithys
Monotropa uniflora
Pyrola asarifolia
Pyrola chlorantha (*P. virens*)
Pyrola elliptica
Pyrola secunda

68. ERICACEAE

Andromeda glaucophylla

Arctostaphylos uva-ursi
Gaultheria hispidula
  (Chiogenes hispidula)
Ledum groenlandicum
Oxycoccus microcarpus
Oxycoccus quadripetalus
Vaccinium caespitosum
Vaccinium myrtilloides
Vaccinium vitis-idaea

69. PRIMULACEAE

Androsace septentrionalis
Lysimachia ciliata
Lysimachia thyrsiflora
Trientalis borealis

70. OLEACEAE

Fraxinus pennsylvanica
  (var. *austinii* and
  var. *subintegerrima*)

71. GENTIANACEAE

Gentiana acuta
  (*G. amarella* ssp. *acuta*)
Gentiana affinis
Gentiana crinita
Gentiana macounii
  (Gentianella crinita
  ssp. *macounii*)
Gentiana rubricaulis
  (*G. linearis*)
Halenia deflexa
Menyanthes trifoliata

72. APOCYNACEAE

Apocynum androsaemifolium

73. ASCLEPIADACEAE

Asclepias ovalifolia
Asclepias speciosa

74. CONVOLVULACEAE

Convolvulus sepium
Cuscuta campestris

75. POLEMONIACEAE

Collomia linearis

76. HYDROPHYLLACEAE

Phacelia franklinii

77. BORAGINACEAE

Hackelia americana
Lappula echinata
Lithospermum canescens
Mertensia paniculata

78. LABIATAE

Agastache foeniculum
Dracocephalum parviflorum
  (Moldavica parviflora)
Dracocephalum thymiflorum
  (Moldavica thymiflora)
Galeopsis tetrahit
Glechoma hederacea
Lycopus americanus
Lycopus asper
Lycopus uniflorus
Mentha arvensis var. *villosa*
Mentha spicata
Monarda fistulosa
Physostegia ledinghamii
  (Dracocephalum nuttallii
  pro parte, *Physostegia*
  parvifolia pro parte)
Prunella vulgaris
Scutellaria galericulata
  var. *pubescens*
Scutellaria lateriflora
Stachys palustris

79. SOLANACEAE

Chamaesaracha grandiflora

80. SCROPHULARIACEAE

Castilleja miniata
Castilleja pallida
   var. *septentrionalis*
Linaria vulgaris
Orthocarpus luteus
Pedicularis lanceolata
Penstemon gracilis
Veronica americana
Veronica comosa
   (var. *glaberrima* and
   var. *glandulosa*)
Veronica peregrina
   var. *xalapensis*
Veronica scutellata

81. LENTIBULARIACEAE

Pinguicula vulgaris
Utricularia minor
Utricularia vulgaris

82. PLANTAGINACEAE

Plantago major

83. RUBIACEAE

Galium aparine
Galium boreale
Galium labradoricum
Galium trifidum
Galium triflorum
Houstonia longifolia

84. CAPRIFOLIACEAE

Diervilla lonicera
Linnaea borealis
   var. *americana*

Lonicera dioica
   var. *glaucescens*
Lonicera involucrata
Lonicera oblongifolia
Lonicera tatarica
Symphoricarpos albus
Symphoricarpos occidentalis
Viburnum edule
Viburnum lentago
Viburnum rafinesquianum
   (*V. affine*)
Viburnum trilobum

85. VALERIANACEAE

Valeriana septentrionalis
   (*V. dioica* ssp. *sylvatica*)

86. CAMPANULACEAE

Campanula aparinoides
Campanula rotundifolia
Campanula uliginosa

87. LOBELIACEAE

Lobelia kalmii

88. COMPOSITAE

Achillea millefolium
Achillea ptarmica
Achillea sibirica
Agoseris glauca
Ambrosia psilostachya
Antennaria campestris
Antennaria neodioica
   (*A. petaloidea*)
Antennaria parvifolia
Arctium minus
Arctium tomentosum
Arnica chamissonis
   ssp. *foliosa*
Arnica cordifolia
Arnica lonchophylla

Artemisia absinthium
Artemisia biennis
Artemisia canadensis
 (*A. caudata*)
Artemisia dracunculus
 (*A. glauca*)
Artemisia frigida
Artemisia ludoviciana
Aster brachyactis
Aster ciliolatus
Aster ericoides
Aster hesperius
Aster junciformis
Aster laevis
Aster puniceus
Aster simplex
Aster umbellatus var. *pubens*
Bidens cernua
Chrysanthemum
 leucanthemum
Cirsium arvense
Cirsium drummondii
Cirsium flodmanii
Cirsium muticum (f. *muticum*
 and f. *lactiflorum*)
Crepis tectorum
Erigeron annuus (*E. ramosus,*
 E. strigosus)
Erigeron asper
Erigeron canadensis
Erigeron elatus
Erigeron glabellus
Erigeron lonchophyllus
Erigeron philadelphicus
Eupatorium maculatum
 (f. *maculatum* and f. *faxoni*)
Gaillardia aristata
Grindelia squarrosa
Helianthus annuus
Helianthus laetiflorus
 var. *subrhomboideus*
Helianthus nuttallii
 (*H. giganteus*)

Helianthus tuberosus
 var. *subcanescens*
Hieracium umbellatum
 (*H. canadense,*
 H. scabriusculum)
Iva xanthifolia
Lactuca biennis
Lactuca pulchella
Liatris ligulistylis
Matricaria maritima
 var. *agrestis*
Matricaria matricarioides
Megalodonta beckii
 (*Bidens beckii*)
Petasites palmatus
Petasites sagittatus
Petasites vitifolius
Prenanthes alba
Prenanthes racemosa
Rudbeckia laciniata
Rudbeckia serotina (*R. hirta*)
Senecio aureus
Senecio congestus
 (*S. palustris*)
Senecio eremophilus
Senecio vulgaris
Solidago bicolor var. *concolor*
 (*S. hispida*)
Solidago canadensis
 (*S. lepida*)
Solidago gigantea
Solidago graminifolia
 var. *major*
Solidago missouriensis
Solidago rigida
Solidago spathulata
Sonchus arvensis
 var. *glabrescens*
Tanacetum vulgare
Taraxacum officinale
Tragopogon dubius

Glossary

acaulescent Apparently without a stem, the leaves and inflorescence arising near the surface of the ground.

achene A small dry indehiscent fruit, distinguished from a nutlet by its relatively thin wall.

acicular Needle-shaped.

acuminate Tapering to a slender point.

acute Forming an acute angle at base or apex.

adnate Grown together or attached; applied only to unlike organs, as stipules adnate to the petiole.

alternate Not opposite to each other on the axis but borne at regular intervals at different levels.

ament A catkin.

androecium A collective term for the stamens.

androgynous (of inflorescence in *Carex*) Denoting a spike that contains both staminate and pistillate flowers, the latter at the base.

annual Of one year's duration.

anther The distal part of a stamen in which pollen is produced, composed usually of two parts known as anther sacs, pollen sacs, or thecae.

anthesis The time during which a flower expands. Often used to designate the flowering period.

antrorse Directed more or less toward the summit of a plant or an organ of a plant.

apetalous Having no petals.

apex Tip.

apical Relating to the apex, or tip.

apiculate   Ending abruptly in a small, usually sharp tip.

appressed   Lying close to or parallel to an organ, as hairs appressed to a leaf or leaves appressed to the stem.

approximate   Close together, but not overlapping.

arachnoid (of pubescence)   Cobwebby; thinly pubescent with relatively long, usually appressed and interlaced hairs.

aril   An appendage growing at or about the hilum of a seed.

aristate   Having an awn, usually terminal in position.

articulate   Jointed; having nodes or joints, or places where separation may naturally take place.

ascending   Growing obliquely upward (of stems); directed obliquely forward with respect to the organ to which they are attached (of parts of a plant).

asexual   Not involving the union of gametes.

attenuate   Gradually tapering to a very slender point.

auricle   A small, ear-shaped projecting lobe or appendage at the base of an organ.

auriculate   Having an auricle.

awn   A slender terminal bristle, usually stiff in proportion to its size.

axil   The angle formed between any two organs.

axillary   Located in or arising from an axis.

axis   The central part of a longitudinal support on which organs or parts are arranged. Compare rachis.

barbed   Provided, usually laterally or marginally, with short reflexed points.

barbellate   Finely barbed.

basal   Located at the base of a plant or of an organ of a plant.

beak   A comparatively short and stout terminal appendage on a thickened organ, such as a seed or a fruit. Not used for a flat organ, such as a leaf.

bearded   Bearing long or stiff hairs.

berry   A fruit developed from a single ovary, fleshy or pulpy throughout, containing one to many seeds; any pulpy or juicy fruit. Compare drupe.

bi- (prefix)   Two, twice, or doubly.

bidentate   Having two teeth.

biennial   Living for 2 years only and blooming in the second year.

bifid   Forked.

bilabiate   Two-lipped.

bipinnate   Doubly or twice pinnate.

bisexual   With both sexes occurring on the same individual. Of flowers, with both stamens and pistils contained in the same flower; a hermaphrodite.

bladder   A modified leaf found on the bladderwort. It is used to trap small aquatic animals.

blade   The expanded part of a flat organ, such as leaf, a petal, or a sepal.

bloom   A whitish powdery and glaucous covering of the surface, often of a waxy nature.

bract   A more-or-less modified leaf subtending a flower or belonging to an inflorescence, or sometimes cauline. Compare spathe.

bracteate, bracted   Having bracts.

bracteolate   Having bractlets.

bracteole   A small bract; a small bract-like organ arising laterally on the pedicel.

bractlet   A bracteole.

branchlet   The ultimate division of a branch.

bristle   A stiff hair, or any slender body, that may be likened to a hog's bristle.

bristly   Provided with bristles.

bud   An undeveloped stem, leaf, or flower.

bud scale   A reduced or specialized leaf that encloses a bud.

bulb   A short, vertical, underground organ for food storage or reproduction on which specialized leaves are prominently developed.

bulblet   A small bulb; usually applied to the bulb-like structures produced by some plants in the axils of the leaves or to structures replacing the flowers.

bur   A hooked fruit.

bush   A shrub.

caducous   Falling off very early.

caespitose   Growing in dense tufts; usually applied only to small plants.

calcareous (of soil)   Rich in lime.

callus   The swollen nodes of the rachilla in Gramineae.

calyx   The outer series of floral leaves that form the perianth of a flower, often green, frequently enclosing the rest of the flower in bud, occasionally colored, petal-like or, in some groups of plants, greatly reduced or completely lacking.

calyx lobe   The free projecting parts of a gamosepalous calyx.

calyx tube   The lower tubular part of a gamosepalous calyx.

campanulate   Bell-shaped.

capillary   Hair-like.

capitate   Head-shaped; collected into a head or a dense cluster.

capsule A dry dehiscent fruit developed from a compound ovary and almost always containing two or more seeds. Compare follicle.

carpel A simple pistil or one member of a compound pistil.

castaneous Chestnut colored.

catkin A dense bracteate spike or raceme bearing many small, naked, or at least apetalous flowers.

caudate Having a slender tail-like terminal appendage.

caulescent Having a well-developed stem above ground.
cauline Situated on or pertaining to the stem.

chaff The receptacular bracts of many species of Compositae.

channeled Grooved longitudinally.

chartaceous Papery in texture.

chlorophyll The green coloring material within the cells of plants.

cilia Marginal hairs.

ciliate Having marginal hairs.

clasping Partly surrounding another organ at the base.

clavate Club-shaped, gradually increasing in diameter toward the summit.

claw The narrow base or stalk of some sepals and petals.

cleft Deeply cut, probably to below the middle. There is no sharp distinction between cleft, lobed (meaning less deeply cut), and parted, (meaning more deeply cut).

columnar Column-shaped or pillar-shaped.

coma A tuft of soft hairs, usually terminal on a seed.

compound (of a leaf) Composed of two or more separate leaflets.

concave Hollowed out; like a saucer.

conduplicate Folded together lengthwise, with the upper surface within, as in the blades of many grasses.

cone A globose to cylindrical arrangement of crowded bracts or scales subtending reproductive organs and usually hard, woody, or long persistent; a structure of similar appearance, although possibly of a different morphological nature.

confluent Flowing or running together.

coniferous Cone-bearing.

connate Grown together or attached. Applied only to like organs, as filaments connate into a tube or leaves connate around the stem. Compare adnate.

convex Having a more-or-less rounded surface.

convolute Rolled up longitudinally; twisted together when in an undeveloped stage.

cordate Heart-shaped; sometimes applied to whole organs, but more often to the base only.

coriaceous Leathery in texture.

corolla The second set of floral leaves of the perianth, often conspicuous by its size or color, but in some plants small and inconspicuous, reduced to nectaries, or lacking.

corymb A type of raceme in which the axis is relatively short and the lower pedicels relatively long, thereby producing a round-topped or flat-topped inflorescence. Sometimes loosely applied to any type of flower cluster of similar shape.

corymbiform Shaped like a corymb.

corymbose In a corymb.

creeping Growing along the surface of the ground and emitting roots at intervals, usually from the nodes.

crenate Dentate with teeth much rounded.

crenulate Finely crenate.

culm The aerial stem of a grass or sedge.

cuneate   Wedge-shaped; narrowly triangular with the acute angle pointed downward.

cupule   A cup-like structure at the base of some fruits (as in some palms) formed by the dry and enlarging floral envelopes.

cusp   A sharp, abrupt, and often rigid point.

cuspidate   Tipped with a cusp or a sharp and firm point.

cyme   A type of inflorescence in which each flower is strictly terminal either to the main axis or to a branch.   See raceme and racemose.   Cymes assume many forms, depending on the number and position of the branches.   They are sometimes distinguished with difficulty from a racemose inflorescence, but may often be determined by the position of the bracts opposite the base of the pedicel instead of below it.

cymose   With the flowers in a cyme.

deciduous   Falling after completion of the normal function; not evergreen.

decompound   More than once compound or divided.

decumbent   Prostrate at base, either erect or ascending elsewhere.

decurrent   Extending downward.   Usually applied to leaves in which the blade is apparently prolonged downward, as two wings along the petiole or the stem.

deflexed   Bent abruptly downward.

dehiscence   The process or act of opening, usually of a fruit.

dehiscent   Opening regularly by valves or slits, as a capsule or anther.

deltoid   Broadly triangular.

dentate   Toothed along the margin, the apex of each tooth sharp and directed outward.

denticulate   Minutely dentate.

dichotomous   Forking more or less regularly into two branches of about equal size.

diffuse Loosely spreading.

digitate Having parts diverging from a common base, as the fingers of a hand. Usually descriptive of leaflets or parts of an inflorescence.

dilate Enlarge.

dimorphic Occurring in two forms.

dioecious Unisexual; bearing staminate and pistillate flowers on separate plants.

disarticulating Separating.

disc An enlargement of or an outgrowth from the receptacle, appearing in the center of the flower of various plants; in Compositae, the central part of the head, composed of tubular flowers.

discoid Resembling a disc; in Compositae, a head composed of tubular flowers only.

distinct Separate; not united; evident.

divergent Inclining away from each other.

divided Cut into distinct parts. Usually describing a leaf cut to the midrib or to the base.

dorsal Located on or pertaining to the back of an organ.

downy Pubescent with soft, fine hairs.

drupe A fleshy or pulpy fruit with the inner portion of the pericarp hard or stony.

drupelet A small drupe, as in a raspberry.

ellipsoid Solid but with an elliptical outline.

elliptical Oval in outline; having narrowed to rounded ends and being widest at or about the middle; of, relating to, or shaped like an ellipse.

endocarp The inner layer of the pericarp, or fruit wall.

entire With a continuous, unbroken margin.

epidermis   The superficial layer of cells.

epiphyte   A plant growing attached to another plant, but not parasitic.

erect   Growing essentially in a vertical position (of a whole plant); describing the position of a structure that extends in the same direction as the organ that bears it (of part of a plant).

erose   Irregularly cut or toothed along the margin.

evergreen   Remaining green throughout the winter.

excurrent   Running out, as a nerve of a leaf projecting beyond the margin.

exfoliate   To come off in scales or flakes.

exsert   To project out or beyond. Often referring to stamens or styles that project beyond the perianth.

exstipulate   Lacking stipules.

fascicle   A small bundle or cluster, without reference to the morphological details of arrangement.

fertile   Capable of normal reproductive functions, as a fertile stamen produces pollen, a fertile pistil produces ovules, a fertile flower normally produces fruit, although it may lack stamens.

fibrous   Resembling fibers.

-fid (suffix)   Deeply cut.

filament   The basal sterile portion of a stamen below the anther, usually slender, sometimes lacking; any thread-like structure.

filiform   Thread-like; long, slender, and terete.

flabellate   Fan-shaped or broadly wedge-shaped.

flaccid   Flabby; lacking in stiffness.

flange   A projecting flat rim or collar.

fleshy   Thick and juicy; succulent.

flexuous   Curved alternately in opposite directions.

floret   A small flower, usually one of several in a cluster.

floricane   The flowering cane, usually the second year's development of the primo-cane, as in *Rubus*.

foliaceous   Leaf-like in flatness, color, and texture.

foliate   Leaved; having leaves.

follicle   A dry dehiscent fruit developed from a simple ovary and dehiscent usually along one suture only.

forked   Divided into nearly equal branches.

free   Not adnate to other organs. Compare distinct.

frond   The expanded, leaf-like portion of a fern.

fruit   The seed-bearing product of a plant, simple, compound, or aggregated, of whatever form.

fulvous   Yellow; tawny.

funnelform   With the tube gradually widening and passing insensibly into the limb.

fusiform   Thick near the middle and tapering at both ends.

gametophyte   In the life cycle, the generation in which sexual organs are produced.

gamopetalous   Having the petals wholly or partly united.

gamosepalous   Having the sepals united.

geniculate   Bent abruptly at the nodes.

glabrate   Becoming glabrous.

glabrous   Lacking pubescence; smooth.

gland   A secreting organ, in plants usually producing nectar or volatile oil and either internal or external.

glandular Containing or bearing glands.

glaucous Gray, grayish green, or bluish green with a thin coat of fine removable particles that are often waxy in texture. Covered or whitened with a bloom.

globose Spherical or nearly so.

globular Spherical.

glomerate Occurring in a compact cluster.

glomerule A compact head-like cyme.

glume One of the two empty chaffy bracts at the base of the spikelet in Gramineae.

glutinous Covered with a sticky substance.

gynaecandrous Having staminate and pistillate flowers in the same spike, the pistillate at the apex.

gynoecium The female portion of the flower. A collective term used for several pistils of a single flower; when only one pistil is present, pistil and gynoecium are synonymous.

gynostegium The structure formed by the union of the androecium and gynoecium in Asclepiadaceae.

habit The general appearance of a plant.

habitat The kind of locality in which a plant grows, such as bogs or woods, for example.

hair An epidermal appendage, usually slender, either simple or variously branched.

haploid Having a single set of unpaired chromosomes.

hastate Having two divergent basal lobes.

head A dense flower-cluster, composed of sessile or nearly sessile flowers crowded on a short axis or disc.

herb A plant, either annual, biennial, or perennial, without a persistent woody stem above ground.

herbaceous Without a persistent woody stem above ground; dying back to the ground at the end of the growing season.

hilum   The scar or point of attachment of the seed.

hirsute   Pubescent with firm, straight, spreading hairs.

hispid   Pubescent with stiff, bristly spreading hairs.

hoary   Grayish-white, close pubescence.

homologous   Similar.

homosporous   Producing spores of only one kind.

hyaline   Translucent or transparent.

hypanthium   An expansion of the receptacle forming a saucer-shaped, cup-shaped, or tubular organ, often simulating a calyx tube and bearing the sepals, the petals, and often the stamens at or near its margin.

imbricate   Overlapping, either in width only, as the sepals or petals of various plants, or in both width and length, as the involucral bracts of many species of Compositae.

immersed   Growing completely under water.

incised   Deeply cut.

incurved   Curved inward.

indehiscent (of fruits)   Not opening at maturity.

indument   Any hairy covering or pubescence.

indurate   Hardened.

indusium   An outgrowth of the frond, wholly or partly covering the sorus in ferns.

inferior   Lower or below.   An inferior ovary is one that is adnate to the hypanthium or to the lower parts of the perianth and therefore appearing to be located below the flower at the summit of the pedicel.

inflorescence   A complete flower cluster, including the axis and bracts.

infrastipular   Below the stipule.

internode   The portion of a stem between one node and the next.

involucel   A secondary involucre.

involucral   Belonging to an involucre.

involucrate   Having an involucre.

involucre   A set of bracts closely associated with each other and subtending an inflorescence.

involute   Rolled inward, so that the lower side of the organ is exposed and the upper concealed. Compare revolute.

irregular (of a flower)   Differing in size, shape, or structure; zygomorphic. Applies to the members of one or more sets of organs (usually the corolla).

keel   A sharp or conspicuous longitudinal ridge; a central dorsal ridge, like the keel of a boat; the two lower united petals in Leguminosae.

labiate   Lipped.

lacerate   Having an irregularly jagged margin; irregularly cut as if torn.

lanate   Woolly.

lance-attenuate   Lanceolate, with the tip tapering.

lance-oblong   Lanceolate and oblong.

lanceolate   Shaped like a lance-head, much longer than wide and widest below the middle.

lateral   Situated on or arising from the side of an organ, as a lateral inflorescence.

leaflet   A single segment of a compound leaf.

legume   A dry fruit derived from a simple ovary and usually dehiscing along two sutures.

lemma   The lower of the two bracts enclosing the flower in Gramineae.

lenticular   Lens-shaped.

ligulate Having a ligule; having the nature of a ligule.

ligule A small, usually flat outgrowth from an organ, as seen at the junction of claw and blade in the petals of some species of Caryophyllaceae or at the junction of sheath and leaf blade in Gramineae; the ligulate corolla of many species of Compositae.

linear Narrow and elongate with parallel sides.

linear-subulate Linear, with an awl-shaped tip.

lingulate Tongue-shaped.

lip Either portion of the limb of a bilabiate corolla or calyx, distinguished as upper lip and lower lip; the odd petal (usually the lowest) in Orchidaceae.

lobe A partial division of an organ such as a leaf. The term generally applies to a division less than half-way to the midrib.

locule A cavity or one of the cavities within an ovary, a fruit, or an anther. The term is often used in preference to the older term, cell.

lyrate Pinnately lobed with the terminal lobe the largest.

macrospore The larger of the two kinds of spores in *Selaginellaceae* and related plants.

malpighian hairs Hairs that are straight and attached by the middle.

marcescent Withering and persistent; usually applied to petals or stamens after anthesis, or to leaves.

mealy Covered with meal or with fine granules.

membranaceous, membranous Thin and pliable, as an ordinary leaf, in contrast to chartaceous, coriaceous, or succulent.

microspore In some pteridophytes, the spore from which the male gametophyte is developed.

midrib The median or central rib of a leaf.

moniliform Resembling a string of beads; constricted at regular intervals.

monoecious Bearing both staminate and pistillate flowers.

mucronate Tipped with a short, sharp, slender point.

multifid Cleft into many lobes or segments.

nectary A gland that secretes nectar, usually on the corolla or disc or within the spur of a flower.

nerve A prominent vein of a leaf or other organ.

nodal Located at or pertaining to a node.

node A point on the stem from which leaves or branches arise; the solid constriction in the culm of a grass.

nodulose Provided with little knots or knobs.

nut A hard, dry, indehiscent, one-seeded fruit or part of a fruit.

nutlet A small nut, loosely distinguished only by its size and scarcely separable from an achene except by the comparative thickness of its wall.

ob- (prefix) In a reverse direction. Usually attached to an adjective indicating shape.

obcordate Heart-shaped, with the point basal.

oblanceolate Lanceolate with the broadest part above the middle.

oblique Slanting; unequal-sided.

oblong Two or three times longer than broad and with nearly parallel sides.

obovate Inverted ovate.

obovoid Having the form of an egg with the broad end apical.

obpyramidal Inversely pyramidal.

obsolete Not evident; rudimentary; extinct.

obtuse Blunt or rounded at the end.

ocrea A sheath around the stem just above the base of a leaf and derived from the stipules. Used chiefly in Polygonaceae.

olivaceous Olive green; olive-colored.

opposite Situated diametrically opposite each other at the same node, as leaves, flowers, or branches; situated directly in front of another organ, as stamens opposite the petals.

orbicular Essentially circular.

oval Broadly elliptical.

ovary The basal, usually expanded portion of a pistil within which the ovules are borne.

ovate Egg-shaped; having an outline like that of an egg, with the broader end basal.

ovoid A solid with an ovate outline.

ovule A reproductive organ within the ovary in which the female structure is produced and that after fertilization becomes a seed.

palea A type of bract in Gramineae.

palmate Having three or more lobes, nerves, leaflets, or branches arising from one point; digitate.

panicle A compound or branched inflorescence of the racemose type. Often applied to any compound inflorescence that is loosely branched and longer than thick.

panicled, paniculate Arranged in a panicle.

papilliform Shaped like a papilla, which is a small, short, blunt, rounded (or cylindrical) projection; nipple-shaped.

papillose (of a surface) Bearing short, blunt, rounded (or cylindrical) projections.

papilonaceous Having differentiated petals (standard, wings, and keel), as in the corolla of many species of Leguminosae.

pappus An outgrowth of hairs, scales, or bristles from the summit of the achene, as occurs in many species of Compositae.

parasite (of a plant) Deriving food and water wholly or chiefly from another plant to which it is attached. Compare epiphyte.

-partite (suffix) Cleft nearly but not quite to the base.

pectinate Pinnatifid into narrow segments of uniform size; comb-like; closely ciliate with comparatively large or stiff and parallel hairs.

pedicel The stalk of a single flower in an inflorescence.

pedicellate Borne on a pedicel.

peduncle The portion of a stem, either leafless or with bracts, that bears an inflorescence or a solitary flower.

pedunculate Having a peduncle.

pendant Hanging down.

pendulous Hanging or drooping.

perennial A plant that continues its growth from year to year.

perfect (of a flower) Having functional stamens and pistils.

perianth The corolla and calyx considered together or either one of them if the other is lacking.

pericarp The wall of a mature ovary; the fruit wall.

perigynium The inflated sac that encloses the ovary in *Carex*.

persistent Remaining attached after the normal function has been completed.

petal A division of the corolla.

petaloid Having the character or appearance of a petal.

petiolate Having a petiole.

petiole The basal stalk-like portion of an ordinary leaf, in contrast with the expanded blade; the support of a leaf.

petiolule The stalk, or petiole, of a leaflet.

phyllary An involucral bract in Compositae.

pilose Having sparse, straight, spreading hairs.

pinna One of the main divisions of a pinnatifid or pinnately compound organ.

pinnate Compound, having branches, lobes, leaflets, or veins arranged on two sides of a rachis.

pinnatifid Having lobes, clefts, or divisions pinnately arranged.

pinnule A secondary pinna; a segment of a bipinnatifid or decompound leaf.

pistil The seed-bearing organ of a flower, consisting of the ovary, style, and stigma.

pistillate Bearing a pistil. Usually applied to flowers that lack stamens.

pith The spongy center of a stem, growing by annual layers.

plumose Feathery. Applied to a slender organ or structure with dense pubescence, such as a style.

pod Strictly, a legume; loosely, often a synonym of capsule.

pollen The spores borne within the anther that produce the male reproductive cells.

polygamous Bearing some perfect and some unisexual flowers.

pome A fleshy fruit, such as an apple or a pear, formed from an inferior ovary with several locules.

precocious Bearing flowers that appear before the leaves.

prickle A small and more or less slender sharp outgrowth from the epidermis.

primocane   The first year's cane (usually without flowers) of *Rubus* and similar genera.

prostrate   Lying flat on the ground.

prothallus   A cellular, usually flat and thallus-like growth, resulting from the germination of a spore, upon which sexual organs and eventually new plants are developed.

puberulent, puberulous   Minutely or sparsely pubescent with scarcely elongate hairs.

pubescence   An indument of hairs, without reference to structure.

pubescent   Bearing hairs on the surface.

punctate   Dotted.   Usually denoting the presence of glands either on the surface or within the tissues.

puncticulate   Minutely punctate.

pyramidal   Pyramid-shaped.

pyriform   Pear-shaped.

raceme   A common type of inflorescence with an elongate unbranched axis and lateral flowers, the lowest opening first. A true raceme is of the racemose type but the term is sometimes loosely applied to a racemiform cyme.

racemose   A general type of inflorescence in which all flowers are axillary and lateral, the axis therefore theoretically capable of indefinite prolongation.   Compare cymose.

rachilla   The rachis of a spikelet in Gramineae and some species of Cyperaceae.

rachis   The axis of an inflorescence or of a compound leaf.

radiate   Spreading from or arranged around a common center.

-ranked (suffix, used with a numerical prefix)   The number of longitudinal rows in which leaves or other structures are arranged along an axis or a rachis.

ray   The ligule or strap-like marginal flower in Compositae.

receptacle   The end of a pedicel or one-flowered peduncle that bears the floral organs; in Compositae, the apex of the peduncle upon which the flowers are inserted.

recurved   Curved downward or backward.

reflexed   Abruptly bent downward or backward.

regular   Describing a flower in which the members of each circle of parts are similar in size and shape.

remote   Scattered; not close together.

reniform   Kidney-shaped; wider than long, rounded in general outline, and with a wide basal sinus.

repent   Creeping or prostrate and rooting at the nodes.

resin   Adhesive substance, insoluble in water, secreted by some plants, e.g., spruce, pine.

resinous   Having resin.

reticulate   In the form of a network.

retrorse   Directed backward or downward.

retuse   Having a small terminal notch in an otherwise rounded or blunt apex.

revolute   Rolled backward, so that the upper surface of the organ is exposed and the lower side more or less concealed.

rhizoid   A single- or several-celled, hair-like structure on the underside of fern prothallia. It functions as an anchor and holds water by capillarity.

rhizomatous   Having a rhizome.

rhizome   An underground usually horizontal stem; a root-stock.

rhombic   Having the outline of an equilateral parallelogram.

rhomboid   A solid with a rhombic outline.

rhomboidal   Having the shape of a rhomboid.

rib   A primary and prominent vein of a leaf.

roseate   Rose-colored.

rosette   A cluster of leaves crowded on very short internodes, often basal in position and circular in form.

rotate   Wheel-shaped. A gamopetalous corolla or gamosepalous calyx widely spreading, without a contracted tube or with only a short and inconspicuous tube.

rudiment   An imperfectly developed and functionally useless organ; a vestige.

runcinate   Sharply incised, with the segments pointing backward.

sagitate   Arrow-shaped; lanceolate or triangular in outline with two retrorse basal lobes.

samara   An indehiscent winged fruit.

scabrous   Rough to the touch, owing to the structure of the epidermis or the presence of short stiff hairs.

scale   Any small thin or flat structure; in Compositae, a single bract of the involucre.

scape   A peduncle with one or more flowers arising directly from the ground or from a very short stem and either leafless or with bracts only.

scapose   Arranged on or borne on a scape.

scarious   Thin, dry, and membranous; not green.

scurfy   Covered with scale-like or bran-like particles.

secund   Directed to one side only, usually by torsion.

seed   A ripened ovule.

sepal   A separate segment of a calyx.

septate   Divided by partitions.

septum   A partition within an organ, as the septa of an ovary or the septa of a leaf in *Juncus*.

sericeous Silky, owing to the presence of numerous soft appressed or ascending hairs.

serrate Toothed along the margin, the apex of each tooth sharp (compare crenate) and directed forward (compare dentate).

serrulate Finely serrate.

sessile Without a stalk of any kind.

setaceous Bristle-like or bristle-shaped.

sheath An organ that wholly or partly surrounds another organ at the base, as the sheathing leaf of a grass.

shrub A woody perennial, smaller than a tree, that usually produces shoots or trunks from the base, not tree-like or with a single bole.

simple (of a pistil) Organized from a single carpel and therefore one-celled, with a single style and stigma. The term is also applied to the ovary alone. A leaf with a single blade, i.e., not compound.

sinuate Having a wavy margin.

sinus The cleft or recess between two lobes.

slough A wet or marshy depression.

sorus (pl. sori) In ferns, a cluster of sporangia.

spadix A form of spike or head with a thick or fleshy axis.

spathe A large, usually solitary bract subtending and often enclosing an inflorescence. The term is used only in the monocotyledons.

spatulate Shaped like a spatula; maintaining its width or somewhat broadened toward the rounded summit; spoon-shaped.

spicate Arranged in a spike.

spiciform Having the form of a spike but not necessarily its technical structure.

spike An elongate inflorescence of the racemose type with sessile or subsessile flowers. The term is often loosely applied to an inflorescence of different morphological nature but of similar superficial appearance.

spikelet A small or secondary spike subtended by a common pair of glumes or bracts, as in Gramineae; in Cyperaceae, a number of empty glumes.

spine A sharp woody or rigid outgrowth from the stem.

spinescent Ending in a spine or bearing a spine.

spinule A small spine.

spinulose Bearing small spines over the surface.

spiral An arrangement of like organs, such as leaves, occurring at regular angular intervals.

sporangium An organ in which spores are produced.

spore A one-celled asexual reproductive cell.

sporophyll A specialized organ for the production of spores in sporangia. Those of flowering plants (pistil, stamen) are often considered to be homologous with leaves; those of gymnosperms and lycopods are the cone scales.

spur A hollow appendage projecting backward from the corolla or the calyx and usually nectarial in function.

squarrose Spreading or recurved at the tip.

stamen A member of the third set of floral organs, typically composed of anther and filament.

staminate Bearing stamens. Usually applied to a flower or plant lacking pistils.

staminode, staminodium A sterile structure occupying the position of a stamen.

stellate Star-shaped. Usually applied to multibranched hairs.

stem A major division of the plant body in contrast to root and leaf, distinguished from both by certain anatomical features and commonly by general aspect.

sterile   Unproductive; infertile.

stigma   The terminal (or by asymmetrical growth occasionally lateral or even basal) portion of a pistil, adapted for the reception and germination of pollen.

stigmatic   Characteristic of or belonging to a stigma.

stipe   The lower part of the petiole, which does not bear pinnae.

stipitate   Having a stipe or stalk.

stipulate   Having stipules.

stipules   A pair of small structures at the base of the petiole of certain leaves, varying from minute to foliaceous and from caducous to persistent.

stolon   A horizontal branch arising at or near the base of a plant, which takes root and develops new plants at the nodes or the apex.

stoloniferous   Producing stolons.

stoma (pl. stomata)   A minute orifice or mouth-like opening between two guard cells in the epidermis, particularly on the lower surface of the leaves, through which gaseous interchange between the atmosphere and the intercellular spaces of the parenchyma is effected.

stone   The hard endocarp of a drupe.

stramineous   Straw-colored.

stranded   Left behind on shore as the water receded.

striate   Marked with fine and usually parallel lines.

strigose   Having appressed, sharp, straight, and stiff hairs pointing in the same direction.

strobile   An inflorescence resembling a spruce or fir cone, partly made up of imbricated bracts or scales.

style   The attenuated part of a pistil that connects the stigma to the ovary.

sub- (prefix)   Slightly; more or less; somewhat.

subopposite   Almost opposite.

subtend   To stand below and close to, as a bract below a flower or a leaf below a bud.

subulate   Awl-shaped.

succulent   Juicy; fleshy.

superior (of an ovary)   Not adnate to other floral organs.

suture   A junction or seam of union; a line of opening or dehiscence.

taproot   Primary descending root.

tendril   A portion of a stem or leaf modified to serve as a holdfast organ.

terete   Circular or essentially so in cross section.

ternate   Arranged in threes.

terrestrial   Growing in the soil, as distinct from growing in water or other habitats.

testa   The outer covering of a seed.

thallus   A plant body not clearly differentiated into stem and leaf, and often without roots or rhizoids.

theca (pl. thecae)   A pollen sac; an anther.

thyrse   A compound inflorescence composed of cymes racemosely arranged.   Also commonly but loosely used to designate a compact panicle.

thyrsiform   Shaped like a thyrse.

tomentose   Woolly, with an indument of crooked matted hairs.

tomentum   An indument of crooked matted hairs.

torus   The receptacle of a flower; in Compositae, the receptacle of the flowers of a head.

trailing   Prostrate but not rooting.

trifoliate, trifoliolate   Having three leaflets.

trigonous   Three-angled.

truncate   Ending abruptly, as if cut off.

tuber   A thickened portion of a rhizome or root, serving for food storage and often for propagation.

tubercle   A small swollen or tuber-like structure, usually distinct in color or texture from the organ on which it is borne, as the tubercle on the achene of *Eleocharis*; a nodule containing bacteria, as on the roots of Leguminosae.

turion   A scaly, often thick and fleshy shoot produced from a bud on an underground rootstock.

tussock   A tuft, mostly used of grasses or grass-like plants.

umbel   A racemose type of inflorescence with a greatly abbreviated axis and elongate pedicels, all arising from one point.   In a compound umbel the branches are again umbellately branched at the summit.

umbellate   Arranged in umbels.

umbellet   One of the small umbels collectively composing a compound umbel.

undulate   Wavy-margined.

uniseriate   Arranged in a single row, series, or layer.

unisexual   Bearing stamens or pistils but not both.

valve   One of the portions of the wall of a capsule into which it separates at dehiscence.   In anthers opening by pores, the portion of the anther wall that covers the pore.

vein   Any of the vascular bundles externally visible in a leaf or other organ, especially those that branch (as distinguished from nerves).

verticil   A whorl of leaves or flowers.

verticillate   Arranged in a whorl.

villose, villous Covered densely with fine long hairs but not matted.

viscid Sticky.

whorl A circle of three or more leaves, branches, or pedicels arising from one node.

wing Any flat structure emerging from the side or summit of an organ; the lateral petals in Leguminosae and Polygalaceae.

zygomorphy The bilateral symmetry exhibited by most irregular flowers, the upper half unlike the lower, the left half a mirror image of the right.

Index

Cody, William J.
Plants of Riding Mountain
National Park, Manitoba

CONVERSION FACTORS

Metric units	Approximate conversion factors	Results in:
LINEAR		
millimetre (mm)	x 0.04	inch
centimetre (cm)	x 0.39	inch
metre (m)	x 3.28	feet
kilometre (km)	x 0.62	mile
AREA		
square centimetre (cm^2)	x 0.15	square inch
square metre (m^2)	x 1.2	square yard
square kilometre (km^2)	x 0.39	square mile
hectare (ha)	x 2.5	acres
VOLUME		
cubic centimetre (cm^3)	x 0.06	cubic inch
cubic metre (m^3)	x 35.31	cubic feet
	x 1.31	cubic yard
CAPACITY		
litre (L)	x 0.035	cubic feet
hectolitre (hL)	x 22	gallons
	x 2.5	bushels
WEIGHT		
gram (g)	x 0.04	oz avdp
kilogram (kg)	x 2.2	lb avdp
tonne (t)	x 1.1	short ton
AGRICULTURAL		
litres per hectare (L/ha)	x 0.089	gallons per acre
	x 0.357	quarts per acre
	x 0.71	pints per acre
millilitres per hectare (mL/ha)	x 0.014	fl. oz per acre
tonnes per hectare (t/ha)	x 0.45	tons per acre
kilograms per hectare (kg/ha)	x 0.89	lb per acre
grams per hectare (g/ha)	x 0.014	oz avdp per acre
plants per hectare (plants/ha)	x 0.405	plants per acre